SATIRE, LIES AND POLITICS

Also by Conal Condren

THE STATUS AND APPRAISAL OF CLASSICAL TEXTS

GEORGE LAWSON'S *POLITICA* AND THE ENGLISH
REVOLUTION

THE LANGUAGE OF POLITICS IN SEVENTEENTH-CENTURY
ENGLAND

Satire, Lies and Politics

The Case of Dr Arbuthnot

Conal Condren
Professor
Department of Political Science
University of New South Wales
Sydney, Australia

First published in Great Britain 1997 by
MACMILLAN PRESS LTD
Houndmills, Basingstoke, Hampshire RG21 6XS and London
Companies and representatives throughout the world

A catalogue record for this book is available from the British Library.

ISBN 0–333–69944–0

First published in the United States of America 1997 by
ST. MARTIN'S PRESS, INC.,
Scholarly and Reference Division,
175 Fifth Avenue, New York, N.Y. 10010

ISBN 0–312–17515–9

Library of Congress Cataloging-in-Publication Data
Condren, Conal.
Satire, lies, and politics : the case of Dr. Arbuthnot / Conal
Condren.
    p.   cm.
Includes bibliographical references and index.
ISBN 0–312–17515–9 (cloth)
1. Arbuthnot, John, 1667–1735—Political and social views.
2. Politics and literature—Great Britain—History—18th century.
3. Arbuthnot, John, 1667–1735. Art of political lying.
4. Political satire, English—History and criticism.
5. Truthfulness and falsehood in literature.   I. Title.
PR3316.A5Z64 1997
824'.5—dc21   97–7121
                                                            CIP

This book is printed on paper suitable for recycling and made from fully managed and
sustained forest sources.

10   9   8   7   6   5   4   3   2   1
06   05   04   03   02   01   00   99   98   97

Printed in Great Britain by
The Ipswich Book Company Ltd
Ipswich, Suffolk

010176

To
H.H.
but for whom and to
W.L.
J.G.A.P.
and
Q.S

Contents

Lying is a language-game that needs to be learned like any other one.

Ludwing Wittgenstein, *Philosophical Investigations*, 249.

Preface

The most curious thing about Dr John Arbuthnot (1667–1735) is just how little studied he has been. He was a successful and enlightened physician, a pre-eminent mathematician, a scientist, an antiquary, a musician and he also achieved fame as a satirist. In this last capacity, his friends Swift, Pope, and Gay held him in the highest regard. Together they formed the Scriblerian Club, a rare meeting of minds bent on satirizing and exposing what they saw as false learning and intellectual pretension. Arbuthnot was to assume a leading role in the Club. Dr Johnson, less than enamoured with the Swiftian circle of satirists, exempted Arbuthnot from his censure and called him a universal genius. Indeed, Arbuthnot's reputation lasted well into the nineteenth-century, one indication of which is the wealth of dubious attributions to his pen. Yet nowadays, and despite the energies of a very large academic community-scrutinizing the Augustan world and its literary output, Arbuthnot is known directly for little more than his shadowy but seminal role in the Scriblerian project and for the probable invention of that satiric image of England, John Bull. What I hope to show in this study is that Arbuthnot's short pamphlet *The Art of Political Lying*, a sham book review printed just after the last of the John Bull tracts and almost on the eve of the formation of the Scriberian Club, is arguably his masterpiece and, regardless of this, a work of such extraordinary density and suggestiveness that it provides a glass in which a whole range of seventeenth and eighteenth-century preoccupations are refracted.

It was Montaigne who remarked that one of the great oddities of civilization concerned the book: invented as the ultimate means of communication, more needed to be written about books than anything else. Arbuthnot's *Art of Political Lying* might seem to have been an exception until this, a study far longer than the text itself. Even worse, what follows is an essay on the nature of the non-existent, surely the ultimate act of academic desperation; for my subject, being an account of the contents of a totally spurious tome, is a reading of a book never written, never intended to be written and presumably never thought to have been written: from the death of the author to the phantom pregnancy of the text. I am aware that I am taking a joke with perilous seriousness. Some willing suspension of belief is thus required of the reader in persisting with what follows, but if given, the re-payment should be a rewarding hermeneutic challenge. Swift suspected in partial admiration that *The Art of Political Lying* would not be understood. I think he can be proven wrong, but even if not, we can at least learn something even by failing.

A friend first introduced me to *The Art of Political Lying* when I was an undergraduate, and supposed to be reading something, I forget what, altogether less interesting and I have carried it around unpacked from my intellectual luggage for some time. Several years ago a colleague and I became involved in teaching a seminar on political satire and parody. Our plan was nothing more cunning than to confront students with a wide range of works called political satire in order to ask what if anything they might have in common and reveal about the political. It was in this context that Arbuthnot's pamphlet assumed a greater significance for me, if not for most of the students who have thought Swift was right about its obscurity. This study, then, is an attempt to delineate the principal contexts and issues it seems to encode and to accept, if only in passing, its invitations to serious reflection. I believe part of Arbuthnot's point in writing the text was for it to act as a stimulus to reflection and so to depart on occasion from strict historicity, may be in the spirit of the work. What follows however, is not a general rumination on a text, a springboard or a mask for sermons on the politics of my own age.

Neither, it must be stressed, does it pretend to be a full account of Arbuthnot and his writings. There is an old but still important biography, G. A. Aitken, *The Life and Works of John Arbuthnot*, (Oxford: Clarendon Press, 1892) written on the eve of Arbuthnot's inexplicable slide into obscurity, and this includes a good range of Arbuthnot's works. There is a fine overview of his intellectual activities, Lester M. Beattie, *John Arbuthnot, Satirist and Mathematician*, (Harvard, 1935 and 1967). There are two later valuable but unpublished doctoral theses, Patricia Carstens, 'Political Satire in the Work of Dr John Arbuthnot' (London, 1958) and Claude Bruneteau, 'John Arbuthnot (1667–1735) et les idées au debut du dix huitieme siècle', (Lille, 1974). There is an fittingly short but acute overview of critical reactions to Arbuthnot by Robert C. Steensma, *Dr John Arbuthnot: Criticism and Interpretation* (New York, Twayne, Macmillan: 1979). There are of course, scholarly works that touch upon him, (Kerby-Miller's on the Scriblerians,

Erskine-Hill's on Pope for example); and there are a few that that deal in detail with aspects of Arbuthnot to one side of my own preoccupation: most notably, Alexander Ross's doctoral thesis, 'The Correspondence of John Arbuthnot' (Cambridge, 1956). Alan Bower and Robert A. Erickson have brilliantly edited Arbuthnot's *History of John Bull* (Oxford: Clarendon Press, 1976), surely a film in the waiting. No doubt I have missed the odd thing. Yet given how dramatically our image of the late seventeenth and early eighteenth centuries has been transformed in almost all respects since most of these works were written, a thorough re-assessment is overdue. A full fresh biography is being planned by Professor Walter O'Grady of Toronto University. What follows is more limited in its scope but if it acts as a spur to a revitalization of interest in Arbuthnot, I will be happy. Why is it that so much academic work is, like that of the politicians, telling other people what to do?

Arbuthnot was, as I have stated, a dauntingly learned man and I have tried to ground what I have to say about his work in traditions of discourse and with reference to texts he knew and owned. This has required cross-reference to his own writings satiric and otherwise, political and scientific. As a result I hope, specifically, to have achieved the following: first, that this study will establish *The Art of Political Lying* as central to Arbuthnot's own *oeuvre*. The extent of this is uncertain, and I have little to contribute to determining its limits. I have remained conservatively within what writers such as Beattie, Ross, Carstens and Bruneteau have accepted. Marginal works, such as *The St Albans Ghost* (1712) may function in my argument just as easily as context as they might as text, although if anything what I have to say on *The St Albans Ghost* strengthens the case for Arbuthnot's authorship or involvement. Similarly, occasional cross-reference between *The Art of Political Lying* and *Peri Bathous, The Art of Sinking in Poetry*, (1727) I believe greatly strengthens the case for seeing Arbuthnot as a proper co-author with Pope (see Appendix B). It is important to stress here, as I shall not labour the point in the text, that I am less concerned with the niceties of specific historical authorship than with an author, a necessary if hypothetical rational agent, and what is being done with authorial and other voices within the text. But issues of specific authorship do become more important in so far as the works more or less confidently attributed to John Arbuthnot form a context for *The Art of Political Lying*, and for which it in turn can become an informing context. It is this matter that is central to Appendix B.

Second, I have given an account of Arbuthnot's satiric style which itself I hope will be seen as a significant counterpoint to that of his friends Swift and Pope, and before them, Dryden, Marvell and Hobbes, and which in conjunction with such writers enriches our understanding of political satire and its limits. Indeed, more broadly my argument would strongly suggest the need for a thorough reconsideration of the relationship between Arbuthnot and his friends Swift, Pope and Gay, their satires and the Scriblerian phenomenon as a whole. Third, I have tried to show how layered and complementary are the intellectual contexts for complex works such as *The Art of Political Lying* which, in so far as they are contextualized, and not just given some background, must have their intellectual identities altered in the process.

Texts may be seen as complex, and I don't necessarily assume complexity to be a virtue, in two principal ways. They may involve intricate argumentation of a specific and relatively homogeneous sort, such as Newton's *Principia*, or Russell and Whitehead *Principia*, or Wittgenstein's *Tractatus* but because of this they evoke fairly simple contexts. Texts however may be complex in terms of the dexterity of argumentation or allusion suggesting, perhaps even within a given sentence complementary contexts. *The Art of Political Lying* is of this second sort. Its simplicity is merely syntactic.

This characteristic of the work has raised a problem of procedure. Providing an abridgement would privilege one sort of context over another, seeing it as mainly an example of this or that; and in any case, the work is sufficiently short for it to be included (see Appendix C). Arbuthnot, I suspect would see the irony in becoming a textual appendix to an attempt to make him intelligible. Further, in the process of elucidation it would have been possible to proceed by dealing *seriatim* with each point as it is made. This however, would have meant elaborating each relevant context over and over again. The alternative I have adopted is to deal with the text as a whole in a series of contexts, effectively re-specifying it and highlighting its different aspects, but on a crucial number of occasions dealing with the same point and its augmented significance in the light of different intellectual surroundings. As lying is the principal subject matter, the disproportionately enlarged and sub-divided context provided by arguments about lying seems justifiable. Some contexts require greater elaboration than others, hence the inelegance of uneven chapter lengths. I have sub-divided each chapter to aid cross-reference, which in itself cuts down on repetition.

The point *inter alia* of this whole study is to re- enforce the view that it tells us little to call anything a text, and that to call something a political text is not much more informative; something which can be gathered cumulatively but only stated baldly here. *The Art of Political Lying* is a work of political theory, of morality, of satire, of parody, of paradox and each predication demands a differing form of framing, and requires seeing it from several perspectives. The difficulty we may have in doing this is partially a function of how the structure of the intellectual world has changed since the early eighteenth century, and has become at times implausibly solidified in the institutionalized structures of university life; thus the reductive absurdity of seeing the work as literature (it is after all satire) and not politics, or vice versa. How much damage the disciplinary division of the humanities has done to historical understanding, often seen to be at their centre, is not taken up here, but *The Art of Political Lying*, is a piece of political rhetoric in quite a precise sense and it is about political rhetoric, in a more general sense and is thus a form of what is now called political theorizing.

I have not attempted a biographical context, though biographical points surface occasionally to serve the case I am making. This may be some relief given the non-existence of that mighty opus, the 'Pseudologia Politicē '. I believe the text we have can only occasionally be highlighted, and then, in no clinching way by available biographical information. I have included a short note (Appendix A) entirely derived from the work of others and which must be seen as particularly provisional in the light of Professor O'Grady's research.

Acknowledgements

It is a pleasure to acknowledge a range of debts. My thanks to Dr Mark Rolfe, for his ear and ideas in trying to teach a seminar on political satire; my thanks also to the smiling forebearence of a number of our students who have allowed Dr Arbuthnot occasionally to displace real political satirists such as Swift, O'Rourke, Jay and Lynn. When I first thought I wanted to find out more about Arbuthnot, I was fortunate to be given the services of John Robertson, involved in a bibliographical project for the School of Information, Library and Archive Studies, UNSW. My thanks to the School and to his fastidious work which was most helpful in plotting the initial lines of this study and to the Library staff of the Humanities and Social Sciences Library at UNSW for their exceptional sense of service. The work was properly conceived and written during 1995–6 when I was a visiting fellow at Churchill College Cambridge. I am grateful to UNSW Special Studies Program Committee for the co-operative way in which leave was granted to me to take up the fellowship. It says much for the atmosphere and surroundings at Churchill that among other ongoing research interests, I could work with such uncharacteristic speed. I owe a considerable debt to the former Master, Professor Alec Broers, and Fellows of the College who displayed fellowship in the most friendly and generous fashion, to the rest of the staff for their kindness and efficiency, and especially to the Vice Master Dr Mark Goldie. Whilst in Cambridge Dr Cathy Curtis, read early drafts of some sections and encouraged me to turn them all into a book. Similar help was given by Dr Andrew Fitzmaurice. I received much stimulation and encouragement from the Tudor and Stuart graduate seminar, conducted by Dr John Morrill, Dr Mark Goldie and Professor Patrick Collinson, to them and to the participants in the seminar my thanks for lively friendship. Professors William Lamont (Sussex) and Iain Hampsher-Monk (Exeter) kindly invited me to give a paper on my work; addressing their seminars helped further focus my ideas. On various occasions Quentin Skinner discussed my work with me and then read the whole penultimate draft with typical acuity and generosity. His recent work on Hobbes which has so obviously helped give substance to one dimension of this study, seems to have been timed for my benefit. The serendipitous circumstances of a singularly Scriblerian discourse on whales gave me the chance to meet Dr Judith Hawley, London University, whose advice on the Scriblerians has been particularly helpful. Dr Murray Simpson of Edinburgh University Library kindly made mss material available to me. Professor Tony Cousins (Macquarie) cheerfully went through the final draft and provided a number of helpful suggestions for which I am grateful. Averil Condren not only suffered me and the good Doctor complicating her own archival work in Cambridge, but volunteered to index the book. The work and vote of confidence are equally appreciated. Naturally, none of of those mentioned here can be held responsible for any way in which Dr Arbuthnot might have misled me. I should like to blame his 'Author' for any blemishes, and as that would be appropriately less than honest, I shall.

A NOTE ON CITATION

There is no adequate uniform issue of Arbuthnot's work. Aitken's *Life and Works* is rare in any case. There are several re-prints of *The Art of Political Lying*, these are not widely available either. Throughout the essay, my references unless otherwise stated are to the first edition of *The Art of Political Lying* printed by John Morphew in 1712, cited in the notes as *APL*, the page numbers refer to those in square brackets in Appendix C.

Part I

1 Introduction

'How will the noble arts of John Overton's *Painting and sculpture now languish! where rich Invention, proper Expression, correct Design, divine Attitudes, and artful Contrast, heighten'd with the Beauties of* Clar-Obscur, *embellish'd thy celebrated Pieces to the Delight and Astonishment of the judicious Multitude! Adieu persuasive Eloquence! the quaint Metaphor, the poinant Irony, the proper Epithet, and the lively Similie, are fled to* Burleigh on the Hill: *Instead of these, we shall have* I know not what –

John Arbuthnot, [*Lewis Baboon Turn'd Honest*. Pref. p. 95]

I

If one imagines a triptych of science, letters and politics in Queen Anne's England, few would have better claims than John Arbuthnot to be depicted on all panels – emphatically to the margins of each. In the world of science and formal learning he was a friend of Newton and of Hans Sloane, a correspondent of the chemist and physician Boerhaave.[1] In 1711 he was court intermediary in the acrimonious exchanges between Newton and Flamstead over the publication of the old astronomer's mathematical calculations.[2] As befitted his status as one of Britain's leading mathematicians, in the following year he helped adjudicate for The Royal Society as to who had primary responsibility for the invention of calculus, Leibnitz or Newton.[3] Yet he was sufficiently casual about his own work for very little of it to have survived him.[4] He was a patron of the young Berkeley who had the highest regard for Arbuthnot's philosophical capacity, although he never did sufficient work to transform talent into achievement.[5] Even his major claim to philosophical fame must re-enforce his marginal status. Contrary to the old encyclopaedias of science, he was not the inventor of mathematical probability theory.[6] A member of The Royal Society and the Royal College of Physicians, all his work on numismatics and diet seem now to be little more than curiosities; his methodologically relentless demolition of Dr Woodward's deluge theory of geology and fossil distribution was a destruction of the insupportable;[7] while the sort of hypotheses he developed late in life about the impact of climate and the air we breathe upon social structure

3

are more readily associated with Montesquieu. A very clubbable and clearly congenial intellectual figure, he was also a founding subscriber of The Royal College of Music, Handel was yet another friend; typically all but one of Arbuthnot's own compositions seems to be lost.[8]

In the amphibious world of court politics at once domestic and international, public and private, Arbuthnot was a political figure by virtue of being the Queen's personal physician. She who ruled to be succeeded was often ill, he in attendance on the ever present cusp of dynastic change.[9] Also a close friend of Robert Harley and of Bolingbroke, even after their rift, Arbuthnot seems never fully to have exploited his status as *eminence grise*.[10] If he did, then in a world that embraced everything from tricky treaties and traduciary pamphlets to whispered bedside counsel and taken tea, he has left too little evidence to push him to the political foreground.[11]

And in that robust and clamorous world of wit and letters, a composition dominated by the egregious Swift and Pope, good Dr John remains a shadowy figure, at the back carrying a spear, whose role in the overall picture of Augustan literary output remains uncertain. His greatest fame has been to carry the title of Pope's *Epistle* to him. In the tomes devoted to Augustan literature, Arbuthnot manages almost invariably a scattering of page references; appearing in the texts like one of Rabbit's Friends and Relations. This is frustrating, for Pope and Swift in particular attest to his playing a central and creative role the evidence can no longer demonstrate. If Lord Chesterfield was right, they failed to do so adequately.[12] The very Arbuthnotian corpus is tumbled with the works of Swift and Pope and has been augmented by unlikely attributions.[13] These at least indicate an eighteenth-century fame analogous to Marvell's satiric infamy in the seventeenth. The name would sell a few copies if attached to anything more adventurous than a stage-coach time-table.[14] As Arbuthnot was careless of his reputation or just plain careless, (perhaps too much is made of this), it is especially difficult to sort the wheat from the chaff.[15] It has proved impossible to discern his precise role in the work of that extraordinary satiric committee the Scriblerians, meeting in his rooms and in the months before the death of The Queen apparently largely under his aegis (see Appendix B).[16]

At least there is the central achievement of the *History of John Bull*, agreed to be Arbuthnot's creation; a work of pace and wit, a remarkably well sustained satiric allegory. It is a Menippean domestic romp on one level which carries great satiric bite by being tied with such decorous ingenuity to the minutiae of politics. So Arbuthnot set on its way a subtle psychologically well-rounded, hence not entirely laudable, cavorting image of England which in transmission gradually became vulgarized to a clear-cut no non-sense emblem of patriotic grit. The John Bull of Gillray's famous scatalogical 'bum boats' image of George III, and even later, the John Bull of *Punch* have little in common but physical girth with 'Gundy Guts', the normally honest, choleric, good-hearted, self-deluding dupe with manslaughtering propensities, that allegorical amalgam of Queen and country, J. Bull cloth merchant to The Old Lord Strutt.[17] The *John Bull* pamphlets which first record the expression 'Law is a Bottomless Pit', appeared in the middle months of 1712. They were mainly a commentary on the issues of peace, money and religious settlement arising from the Scottish Union and above all, the debilitating Wars of the Spanish Succession, with their uncomfortable alliances and shifting antagonisms. These pamphlets, as Swift put it, revealed a deep seam of satiric genius in Arbuthnot and established his literary fame.[18]

Yet in October, shortly after the appearance of the final part, *Lewis Baboon turned Honest*, while looking after his own sick family as well as attending the quasi-queezy Queen at Windsor, Arbuthnot sent the manuscript of *The Art of Political Lying* to Swift. He in turn, immediately sent for the printer Morphew and shortly after it was advertised in *The Examiner*.[19] This brief satire stands to Arbuthnot's corpus as the man himself stands to the larger panels of Augustan England, brilliant but marginal. Swift himself called it a very pretty piece, and some have thought it too good for Dr John. Indeed, had it been written by Swift, I think there is little doubt it would have become more central to literary scholarship. Often referred to in passing, it has been little discussed or analysed. Aitken concluded that despite its gravity and wit it did not have the interest of *John Bull*.[20] Yet later he seems to have changed his mind calling it a 'delightful skit' worthy of Swift at his best, doing nothing to make good the claim.[21] George Kitchin noted only that Arbuthnot originated the parody of the book review.[22] Hugh Walker, who regarded the wit of Arbuthnot's 'fugitive writings . . . singularly brilliant', called it his masterpiece, but also left it at that.[23] Fifty years later Patricia Carstens concurred, but her fine doctoral thesis was not published, a non-event which itself has helped keep *The Art of Political Lying* more peripheral than it

deserves.[24] E. L. Steeves quite overlooks it, to the detriment of her
analysis of the authorship of *Peri Bathous, or The Art of Sinking in
Poetry*; the same oversight mars Attilio Brilli's otherwise excellent
analysis of the work and its immediate background.[25] In his
similarly valuable study of Scriblerian writing as a satiric type,
Brean Hammond makes no mention of what on his own criteria, is
a fundamentally Scriblerian piece.[26] Kerby-Miller, however, passingly notes that *The Art of Political Lying* might form part of the
literary context for *Peri Bathous*, while Claude Bruneteau tantalizingly calls the *Art of Political Lying* a prelude to the whole
Scriblerian satiric enterprise but focuses disproportionately little of
his attention on the very satire he seems to suggest should be seen
as so central.[27] As the Scriblerian Club was formed and set about
its project (the word is chosen with care)[28] after the first printing of
The Art of Political Lying, my discussion of that work's relationship
to Scriberlian satire in general and the *Peri Bathous* in particular is
mainly reserved for Appendix B. In the preceeding bulk of my
argument, the point is simply to show that *The Art of Political
Lying* is too interesting to have been marginalised as it has.
Although I cannot with confidence explain why this has been so, an
hypothesis will itself help in an initial characterisation of the work.

Our expectations of satire have been strongly coloured by the
theories and practices of moderns such as Dryden, Swift and Pope
elaborating on the ancients of Rome, especially Horace and Juvenal.[29] If we expect late seventeenth and early eighteenth-century
satire to be well crafted, we also expect the obvious, if posturing,
occupation of the high moral ground. And we anticipate finding not
a little of the victim's blood beneath the satirist's feet; thus for
example, the recent rediscovery that *Leviathan* stands in part as a
powerful work of satire has been dependent upon nosing Hobbes's
most wounded targets. Apart from its polished and simple style,
however, all these expectations will be dashed by *The Art of
Political Lying*. If we look more to a Menippean model of satire,
there must be a shock of alienation from something that follows so
hard on the coat-tails of *John Bull*. There is in it nothing of the
bracing slippage between satire and diatribe and insinuated slander,
no anger, no defamation or derision, no bald insult; the moral edge
always maintained by the satiric persona seems dulled or only
intermittently visible. The whole of the little work is cool, indirect,
ironic and elegant. So, it has been apt to slide past with little more
than nods of appreciation at how well its central conceit is

sustained. Formally, *The Art of Political Lying* is an account of what someone else has in press, the systematic *summa* 'Pseudologiā politicē', for which it is a judiciously balanced 'puff'. Even in the context of *John Bull*, this may look odd. By Pope's or Juvenal's standards Arbuthnot no doubt pulls his punches in *John Bull*, but Marlborough (alias Humphrey Hocus) and the Godolphin ministry (the first Mrs Bull) are bloodied nonetheless. There is no regret when John kills his wife in a domestic tiff. Heroes might well be flawed and villains like Nick Frog have some almost saving graces but they remain unequivocally heroes and villains. Decode the names and the message and moral censure are clear. There is nothing so straightforward in the *Art of Political Lying*. At the time Swift indicated that this 'very pretty piece' was not so obvious as to be understood, thus intimating that the allusions and perhaps above all the satiric focus would be hard to grasp.[30] Had it established a more central place in a satiric canon, it would have been more difficult for Alvin Kernan to assert that as a species satirists have little respect for the intelligence of their audience.[31] This, as Swift knew, is satire by stealth. Thus it may seem too bland for any offence, or even to carry much point, it slips by because Arbuthnot may have assumed too much of his audience's wit.

II

If our expectations about satire may have helped keep *The Art of Political Lying* peripheral to the study of Augustan writing, it will be necessary in what follows to reconstruct Arbuthnot's understanding of satire. That it is not adequately contextualized by reference to Marvell's practice and Dryden's theory[32] and that it is different from Swift's and Pope's will become clear. Indeed, Pope's *Epistle to Dr Arbuthnot* and the collapse of their joint satiric project *The Art of Sinking in Poetry* help draw up discernible, if retrospective lines of demarcation.[33] What I shall suggest is that *The Art of Political Lying* exemplifies a view of satire at once rhetorically informed and sophisticated, which adds to our understanding of the satiric impulse and the problematic status of the satirist in society. The issues that Pope took up in his *Epistle to Dr Arbuthnot* are not one-sided even if, formally, we have had only one side of the argument. Pope's aggressive justification for his personal satire can almost be seen as a response to the satiric style of the *Art of Political Lying*, which itself may be seen in the context of Swift's famous remark that satire is like a glass wherein we behold everyone's face but our own.[34]

Lying also may have proved to be a problem in taking Arbuthnot's work as seriously as we might. Too easily we fall foul of the amused or contemptuous truism that all politicians are liars. As John Barnes remarks, citing Arbuthnot in the company of Hannah Arendt, in the political domain lies are expected almost as a matter of course.[35] Arbuthnot might be on our side, as it were, but that he is descanting only upon the universally applicable is a dangerously slick conclusion.[36] Even the careful and appreciative Patricia Carstens is persuaded that Arbuthnot is showing us how the world really works.[37] He is not. The central *topos* of lying will need much more attention and this will show Arbuthnot to be establishing an economy of political discourse which is subtle, serious and provoking.

For the remainder of this essay, I propose to look at the work afresh, treating it as the many faceted jewel in the Arbuthnotian satiric crown. This will require that I move across a broad range of his works so showing *The Art of Political Lying* to be a concentration of his preoccupations and intellectual characteristics. By this token, any single context does less than justice to the issues the work entraps. In catching the light of a number of overlapping intellectual contexts, that 'pretty piece' makes John Arbuthnot an untoward point of focus in the chiaro-scuro images of the Stuart dusk.

To an extent *The Art of Political Lying* is itself a somewhat disingenuous commentary on contemporary and party generated accusations of dishonesty. This topical dimension to the work, however, is the means by which more long-standing problems and intellectual traditions are brought into conjunction. It is an oblique parody of views on theoretical knowledge, which elsewhere, in his scientific work, Arbuthnot discussed directly and seriously. More specifically, it provides a parody of rhetoric and its ancient claims to be the fitting theoretical knowledge for politics. It is an ironic aside on the ancients and moderns debate. It is an inversion of the propensities of casuistry which had taken more than a century to work-themselves out. Again, more specifically, it is a satire of Machiavellian and post-Machiavellian political theories; and it is a comment on contemporary changes in the lexicon of English political values, on the fraught and refracted relationships between the terms of public virtue and private worth. It offers a glimpse of a clear and consistent picture of an entirely secular political system fashioned by Arbuthnot's enhanced sense of metaphorical decorum. *The Art of Political Lying* shows that same capacity to generate a model of understanding through figurative language that marks the *John Bull* satires. In addition, however, it also shows, as I have noted, a style of satire at odds with the earlier work and which is fully justifiable in terms of the rhetorical theories which run distorted through the whole. And if this claim seems implausibly reflexive, we might end by noting the overlooked yet most obvious feature of the pamphlet. It is a lie about lying and so, but for Lucian's *Verae historiae* and Oscar Wilde's much later Lucianic tribute, *The True Art of Lying*, it is perhaps the most elaborate example we have of that class of formal paradoxes, *insolubilia*, the paradox of the liar. It was crafted by one who liked nothing better than to satirize the claims of philosophy and science, and thereby, like Lucian, Arbuthnot would say something serious, even enticing us to the limits of what we think we know. The study will begin by looking at the work's topical references and personal allusions as these lead directly to the sort of satiric theory in and behind *The Art of Political Lying*. It will end with a discussion of the tract as a paradox, and between, take up the other principal topics it suggests.

NOTES

1 On Herman Boerhaave see, John P. Wright, 'Boerhaave on Minds, Human Beings and Mental Diseases', *Studies in Eighteenth-Century Culture*, 20, (1990).

2 George A. Aitken, *The Life and Works of John Arbuthnot*, (Oxford: Clarendon Press, 1892), pp. 36–37: Alexander Ross, 'The Correspondence of John Arbuthnot', Cambridge University, unpublished PhD thesis, 1956, vol 1. p. 44.

3 Aitken, *Life and Works*, p. 37; a good account of this and the Flamstead controversy is in Lester M. Beattie, *John Arbuthnot mathematician and satirist*, (Camb. Mass.: Harvard University Press, 1935; and New York: Russell & Russell, 1967). The flurry of sometimes intemperate letters involving Newton, Flamstead, Halley, Arbuthnot are in Ross, 'Correspondence' vol. 1.

4 Two small (conflated) pieces on probability survive (1690 and c.1696), both dealing with issues germane to his translation of Huygens *De rationiis in ludo aleae, The Laws of Chance*, (1694) with the longer ms (c.1696) also anticipating his work on child mortality, see Gregory Mss (Edinburgh University) Dk. 1.2^2 Fol. B no. 19.

5 Aitken, *Life and Works*, pp. 54–5 quoting Berkeley (Sloane MSS 4036, Fol. 167); see also Christopher Fox, *Locke and the Scriblerians*, (Los Angeles: University of California Press, 1988), p. 19, citing Swift, *Correspondence* ed. Harold Williams, (Oxford: Clarendon Press, 1963), vol. 2; Fox also notes Dugald Stewart's belief that for all his lightness, the intelligent reader will see a metaphysical depth and soundness in Arbuthnot p. 95.

6 John Arbuthnot, *An Argument for Divine Providence, taken from the constant regularity observed in the birth of both sexes*, Philosophical Transactions of the Royal Society,1710, vol. 27; *The Dictionary of Scientific Biography*, (New York: Scribners, 1970), vol. 1. p. 208 attributes to this work the origins of mathematical statistics.

7 John Arbuthnot, *An Examination of Dr Woodward's Account of the Deluge*, (1697).

8 Patricia Carstens, 'Political Satire in the Work of John Arbuthnot' London University unpublished PhD thesis, (1958), p. 3 where she also notes he was a Freemason. For details of Arbuthnot's involvement and role in early eighteenth-century music, Claude Bruneteau, 'John Arbuthnot (1667–1735) et les idées au debut du dix huitieme siècle', Doctoral Thesis, Universite de Lille (1974), 2 vols, pt. 1. ch. 2., pp. 86–93. Samuel Johnson, *Lives of the English Poets*, ed. George B. Hill, (New York: Octagon 1967 edn.) p. 273 refers to one surviving anthem, 'As pants the heart', in the books of The Chapel Royal.

9 The quip about succession is usually attributed to G. M. Trevelyan.

10 Aitken, *Life and Works*, pp. 45–7, 71ff.

11 Jonathan Swift, *Journal to Stella*, ed. Harold Williams, (Oxford: Clarendon Press, 1948), 26, Sept. 1711; Ross,'Correspondence' vol. 1, 37–8.

12 Lord Chesterfield, *Letters*, an accusation aired by Dr Johnson, *Lives of the Poets*, 3. p. 273.

13 For discussions of authorship which have by and large pared down his *oeuvre* see Beattie, *John Arbuthnot;* Robert C. Steensma, *Dr John Arbuthnot: Criticism and Interpretation, (New York: Twayne, 1979);* Patricia Carstens, 'Political Satire', has passing comments throughout.

14 A further indication of his established fame is in the number of citations in Noah Webster's first dictionary, where the people quoted to illustrate meanings were ones likely to be familiar to and carry authority with Americans. I am grateful to David Mikhosz for this point.

15 The issue is not helped by Arbuthnot's son disowning the whole posthumous edition of Arbuthnot's works, *Miscellaneous Works*, (Glasgow, 1751), 2 vols., though it contains material that has always been accepted as certainly his. Such issues are not crucial to the argument of this book.

16   See Beattie, *John Arbuthnot;* Christopher Fox, *Locke and the Scriblerians* pp. *10–19*; and especially Charles Kerby-Miller ed., *The Memoirs of the Extraordinary Life, Works and Discoveries of Martinus Scriblerus* (Oxford: University Press, 1988), Introduction, at length.

17   All references are to the splendid *History of John Bull*, eds., Alan W. Bower and Robert A Erickson, (Oxford: Clarendon Press, 1976).

18   For the best discussion, Bower and Erickson, 'Introduction' to *History of John Bull.*

19   Ross 'Correspondence', vol. 1, p. 225 Arbuthnot to Sloane, 4, Aug., 1712 (Sloane MSS 4043, f. 76); Swift, *Journal to Stella,* 9, Sept., 1712; Aitken, *Life and Works*, p. 51.

20   Aitken, *Life and Works,* p. 51.

21   Aitken, 'Arbuthnot and Lesser Prose Writers', *The Cambridge History of English Literature*, ed. A. W. Ward and A. R. Waller, (Cambridge: University Press, 1912), vol. 9 p. 136.

22   George Kitchin, *A Survey of Burlesque and Parody in English*, (1931, reprinted New York: Russell and Russell, 1967), p. 159.

23   Hugh Walker, *The English Essay and Essayists*, (London: Dent, 1915), p. 126; English Satire and Satirists, (Dent: London, 1925 reprinted Octagon, New York, 1965) repeats the praise but only summarizes the work, p. 199ff.

24   Carstens, 'Political Satire', ch. 6, p. 325; followed by Bruneteau,'John Arbuthnot et les idées' pt. 4. ch. 2, p. 683.

25   Attilio Brilli, *Retorica della satira, con il Peri Bathous, o L'arte d'inchinarsi in poesia di Martinus Scriblerus,* (Bologna: Il Mulino, 1973).

26   Brean Hammond, 'Scriblerian Self-Fashioning', *The Year Book of English Studies*, 18 (1988), pp. 108–24.

27   *The Art of Sinking in Poetry: Martinus Scriblerus' Peri Bathous, A Critical Edition*, ed. E. L. Steeves, (New York: Columbia University, Crown Press, 1952), Introduction; Kerby-Miller ed. *Memoirs*, Introduction p. 72; Bruneteau, 'John Arbuthnot et les idées', pt. 1. ch. 2, p. 59; see below Appendix B. What Bruneteau says specifically of *The Art of Political Lying*, largely, if soundly follows Carstens and Beattie.

28   This is to disassociate myself from such expressions as the 'Enlightenment project' as anachronistic and misleading. In contrast the Scriberlian Club was the organisation of a satiric project only partially carried out.

29   Howard Erskine-Hill, *The Augustan Idea in English Literature*, (London: Edward Arnold, 1983) for a fine account of the ambivalent inspiration of Rome.

30   Swift, *Journal to Stella*, 12, Dec., 1712; Aitken, *Life and Works*, p. 52.

31   Alvin Kernan, *The Plot of Satire*, (Yale: New Haven, 1965), p. 24.

32   Carstens, 'Political Satire', ch. 2; also more generally Brilli, *Retorica della satirica*, who sees Dryden's theory of satire as the direct antecedent of all Scriberian satiric theory, p. 28.

33   See below ch. 3 VI; Appendix B.

34   Swift, *The Battle of the Books*, (1710) Preface.
35   John Barnes, *A Pack of Lies: Towards a Sociology of Lying*, (Cambridge: University Press, 1994), pp. 30–1.
36   A. F. Pollard, *Political Pamphlets* (London: Kegan Paul, 1897), p. 105, for a coy intimation of this; See also James Sutherland, *English Satire* (Cambridge: University Press, 1962), p. 101; Beattie, *John Arbuthnot*, p. 298.
37   Carstens, 'Political Satire', ch. 6. p. 297ff.

2 Topicality and the Limits of Party Commitment

The Doctrine of unlimited Chastity and fidelity in Wives, was universally espous'd by all Husbands, who went about the Country, and made the Wives sign Papers, signifying their utter Detestation and Abhorrence of Mrs. *Bull's* wicked Doctrine of the indispensible Duty of Cuckoldom. Some yielded, others refused to part with their native Liberty; which gave rise to two great Parties amongst the Wives, the *Devoto's* and the *Hitts*. Tho' it must be own'd, the distinction was more nominal than real; for the *Devoto's* would abuse Freedoms sometimes; and those who were distinguish'd by the Name of *Hitts*, were often very honest.

John Arbuthnot, *John Bull in his Senses*, ch. 2.

I

There is more than a dynastic continuity in Stuart politics. The Civil Wars cast a shadow of insecurity and suspicion well into the eighteenth century. Even if, retrospectively, the fear of war might sometimes seem to have been orchestrated and operatic, those years of blood, followed by insurrection and conspiracy in every decade for the next seventy years sustained distrust. Queen Anne lived and died in the long seventeenth century, expiring just before a further if truncated attempt to recommence the British Civil Wars. The politically excluded or marginal orders may well have been reasonably accepting of their lot but from any evidence of this one cannot infer an established political stability. Unrest does not have to come from below and seventeenth-century experience strongly suggested that the political nation itself was a prime source of intermittently organized instability. Anne's reign was understandably then, a period in which the view was still entrenched that any faction, or party was itself destabilizing, was an unsettling interest working against the public good; thus group identity was very much itself a part of the ammunition of public accusation. There was always an easy slippage from accusations of disingenuous self-interest to mendacious conspiracy, treachery and rebellious intent. Yet to cope with other, imagined, or perceived factions, men had to form alliances, organize themselves into what their enemies in turn would see as a party. Even anarchists organize themselves. The necessity to associate with a party to combat factional interest was morally compromising, a choice, as Swift maintained, between less than entirely laudable alternatives.[1]

Similarly, party labels were easily generated, variously applied, appropriated or subverted, (Tory, Old Whig, New Whig, Whig) with a cavalier disregard for the difficulties this would cause later historians.[2] The situation is exacerbated by the fact that London played host, as it were, to national politics, which was thus superimposed on a complex, jealously guarded and tensile metropolitan politics.[3] A monarchy sat crowned, upon a petty republic. Whatever else they might have reflected, the labels Whig and Tory were used to express widespread and diverse fears of deep and dangerous divisions in the political nation. More formally, they also corresponded to voting patterns in Parliament.[4] Yet often allegiances were personal, shifting or apparently eccentric. Thus while men could readily enough see enemies as belonging to parties, they would style themselves as allied with friends or neighbours. In these ways, the existence of parties does not suggest established and traditional structures appropriate to later piping times of peace.[5] Indeed, even the more ameliorating accounts of party conflict were apt to trace their origins to the Civil Wars and so enshrine those disturbances in the present.[6]

But whatever their parliamentary, organizational and nascent financial stability, it is in the public world of print that the Whigs and Tories of Anne's reign survive most vividly. Around these names coagulated a remarkable eristic literature. The beginnings of this burgeoning of debate are sometimes dated from 1695, when governmental control over the press was abandoned. But censorship had never been more than partially effective and attempts to control the press were to continue.[7] In the present context, it is more important to put aside the symbolic associations of 1695 as initiating a newly developing 'public sphere' of free-flowing information and rational debate.[8] That notion functions best as a myth of consolation when contemplating the power of twentieth-century communications technology. It may even be that public debate was damaged by the sheer bulk of the printed word. What issued from the presses was seen even at the time as a bewildering and promiscuous flow of news, rumour, gossip, misinformation, advertisement, promotion and opinion.[9] The effect was to aggravate, or even for dramatic effect, to help create social divisions most conveniently organized by the deployment of 'Whig' and 'Tory'.[10] These terms, if they did not reflect deep and simple fissures in society, expressed symbolically distrust and what was perceived to be irrational hostility. Most who wrote for a party proved the point by uncompromising and alienating attack on stereotyped adversaries.[11] Rather than functioning as centres for the studied appraisal of the news in general, coffee houses may well have reinforced prejudice and division, or acted to aid the exchange of specialised information in a narrow vortex.[12]

II

Polemical writing was no newer in the eighteenth century than the coffee house. The Civil Wars had seen an extraordinary efflorescence of the printed word and established a view to be shared by many throughout the Interregnum and Stuart era that rational discourse was smothered by group interest. Affording some plausibility to this fear, there had been sporadic alliances between letters and propaganda from the end of the sixteenth century and more recently in the publication of Dryden's *Absolem and Achitophel.* What is witnessed from 1695, however, is an increasing need to mobilize the printed word for the promotion of policy and the co-ordination of parliamentary support.[13] None understood this better than Arbuthnot's close friend Robert Harley and he developed with Bolingbroke a remarkable propaganda effort to promote peace with France.[14] Geoffrey Holmes has claimed that the Wars of the Spanish Succession had provided the greatest agent for party division, especially when from 1710 peace became an increasingly practical solution.[15] When Harley formally came to power through an unusually dramatic change of government in October of that year, he was determined to achieve peace and clearly saw party extremity as a barrier. Hostile to implacable war-party Whigs, he did not want to rely exclusively upon similarly extreme Tories, and so tried to found his support initially upon a 'broad bottom'.[16] His public manifesto to this effect, *Faults on Both Sides* itself had generated no little distrust among those expecting the maintenance of hostilities.[17]

Swift bent his energies to Harley's needs and we can see in the publication of his *Sentiments of a Church of England Man,* 1711 the irony of a Tory interest in appealing across party division.[18] Irrespective of the motive for the publication of this earlier tract, it does point to what were perceived throughout the first decade of the century as the debilitating dangers of party affiliation. In an unusually strenuous effort to maintain balance Swift remarked that each party was automatically armed with accusations that its adversaries were not to be believed.[19] Neither party, he continued, really made enquiries after truth but exercised itself with 'injurious appellations', accusing their enemies of holding 'horrid opinions' and then reproaching them for lack of charity.[20] The result was to maintain an ethos of extremity often aided by the evocation of seventeenth-century divisions as a direct lineage for present discontents.[21] The rhetoric of moderation remained, as it had been largely in the previous century, less a location between extremes than a claim on integrity in the face of an extreme. It was, remarked Swift, a word almost entirely in the hands of bigots.[22] As Charles Davenant had also put it, moderation was as 'specious' as it was 'desirable'.[23]

Despite the projection of a Habermasian public sphere onto the early eighteenth century, diverse political antagonisms conjoined with uninhibited ephemeral printing meant that hefty portions of argument could not hope to satisfy innocent expectations of rational persuasion on the basis of adequate data equitably shared (see below ch. 7, VI). As Swift suggested, in a passage of exasperated exaggeration, debate exhibited mindless allegiance rather than judgement; 'entering a party' was like joining an order of friars and was 'unsuitable to civil and religious liberties'.[24] But if Swift overstates the rigours of party solidarity, his fears about the nature of debate seem more sustainable.[25] As I shall suggest, his image of mindless and defensive antagonism to adversaries helps set out a central problem of public debate to which *The Art of Political Lying* was a considered solution. In the meantime, we may say at least that public political discourse was preoccupied with mobilising, reassuring and preaching to the converted, shoring up the waverers, gathering the fugitive 'party'. For as Davenant had remarked, 'all factions are many headed and heads look in different ways'.[26] And in 1712, a year of extreme and clamorous dispute,[27] the victorious Tories were already looking in different ways and transmogrifying into a self-destructive hydra.[28] Even now the consolidatory functions of rhetoric are not inconsiderable.

III

With these general comments in mind, what does it mean to call *The Art of Political Lying* a Tory tract and how might it be seen as a response to the problems of political debate articulated by writers such as Swift? Arbuthnot was an episcopalian friend of Swift, Harley and Bolingbroke, whose work was printed by Morphew, the principal printer of Tory propaganda. He was very much a court figure. His brothers were Jacobites and were to live in exile after 1715 and he was to keep in familial touch with them. In 1712 Arbuthnot was hostile to Godolphin, Churchill, Sunderland, Nottingham and any who might be seen as part of a war party, as *John Bull*, a direct sally on Harley's behalf, makes amply clear. Only a few years before, however, when Harley was in an uncertain coalition ministry, Arbuthnot along with Defoe was aligned with it also over the unification with Scotland,[29] – all on the side of Her Majesty who feared that without unification, the parties and factions around her would harden into civil war camps after her death, no poor judgement.

To recognize *The Art of Political Lying* as a Tory tract however, is not to see it as expressing an ideology, or a platform, or set of policies appropriate to a formal disciplined organisation designed for the purpose of government. Such an 'ideology', if the term is appropriate, provides a less than comfy fit for all those who might have worn the Tory label. To call this satire Tory, however, is at least to place it in a context of a general attitude of intense loyalty to the established Church, supported by belief in peace with France, usually expressed through ossifying rhetorics of accusation and defence against divisiveness, faction, corrosion of the public interest and of course lying. More specifically, *The Art of Political Lying* draws on and ironically generalizes from the way in which mutual accusations about lying, cheating, and misinformation played their part in maintaining, or even creating group identity. It was almost a presupposition of debate that opposing parties were always dishonest, dishonesty helped explain why they existed. Most generally characterized, Arbuthnot's satire exposed this common belief by pretending to take it with deadly literalness.

Because it was a satiric allegory, *John Bull* had not formally divided the world into Whigs and Tories, but those who wanted to press the law suit (Whigs) and those who wanted a reasonable settlement (Tories) and it had been tied in the most intimate detail to peace and war parties in ways that demonstrated clear unswerving commitment. *The Art of Political Lying*, however, does directly divide the world into Whig or Tory. Yet it also displays a degree of moderation and a rational, judicious balance which would have been appropriate to Harley's strategies of broadening the base of his support, endeavours which were proving hazardous and erratic in the course of 1712. One may tentatively go further. Harley was genuinely suspicious of parties and did not fit any simple party mould and tried as far as he could throughout 1712 to maintain a mixed administration, bringing increasing tensions both with Bolingbroke and with his back bench country Tory support.[30] Given that people professed to be against party and for moderation, it is superficially curious that his ameliorating *Faults on Both Sides* generated its own controversy, being attacked from both sides.[31] This is because Harley's own explanation of party faction was of a neo-Harringtonian, almost (to use another dangerous word,) republican nature. Lord Chesterfield remarked of Arbuthnot that by prejudice he was a Jacobite, by reason a republican.[32] Epigram may sacrifice some accuracy, but what it supports is that throughout 1712 Arbuthnot was a Harleian tory. The only point of tension possibly concerned Scotland. In *John Bull* Sir Roger Bold (Harley) is portrayed effectively as being a party to the betrayal of Jack (Scottish Presbyterianism).' Then let him hang . . . the Fellow has been mad these twenty Years. With this he slunk away.'[33] And Arbuthnot may have looked with ambivalence on Harley's attempts to slow the passage of the Episcopalian Toleration Bill in order to maintain good relations with the Scottish Kirk. Be this as it may, Arbuthnot had and retained sufficiently intimate relations with Harley (soon to become a fellow Scriblerian) to suggest that the *Art of Political Lying* was consistent with Harley's strategy of government. Such a consistency required in its turn inclusive rather than exclusive strategies of argument.

IV

Having spoken generally of *The Art of Political Lying*, at this point it is necessary to emphasize the importance of distinguishing its personae. The pamphlet purports to be an account of '*Pseudologia politicē*', a learned two volume book in press written by an unnamed Author and which is presented to us only through its

publicist. The sustained personae of the book's Author and Projector and their textual interaction allows Arbuthnot to maintain an
ironic distance from his own creation; they become the means by
which he can display opinion, prejudice and commitment appropriate to the topics and visions of the day.

As Patricia Carstens has pointed out, pretty well all the topical
allusions and examples of lying that are reported from the 'Pseudologia' by the Projector, reflect accusations against Whigs or subvert
Whig fears.[34] Indeed, she goes so far as to suggest that both the
Author of the 'Pseudologia' and his Projector are mildly whiggish,
forming an alliance of pedantry and Grub Street.[35] The blandness
with which allusive accusations are presented indicates either that
the Author of the 'Pseudologia' or the persona of the puff for it
take accusations as simple truths. If they are truly 'Whigs' it would
seem to follow at first blush that Arbuthnot is satirizing according
to all the established expectations of party animosity by parading
their folly and barefaced acceptance of dishonesty. This is certainly
Carstens' view. Understanding the satire requires we identify the
targets and this leads her to a sort of whiggish reductionism, an
entrapment in party dichotomies from which Arbuthnot was more
free.[36] Nevertheless, we can get some way towards identifying the
most immediate if not the exclusive targets: the examples of and
advice concerning *tō phōberon*, the terrifying lie, certainly reflect
Whig and war party accusatory rhetoric and Tory belief that such
fears were orchestrated.[37] The King of France and the Pretender
must be brought out once a year, but not too often or they will
simply cease to scare people.[38] With respect to keeping lies within
the bounds of plausibility, Carstens is right to say that John
Churchill is probably the intended victim of the advice concerning
'additory lies', those lies that inflate a man's reputation. Churchill
was widely renowned for acquisitiveness and the 'Pseudologia' is
reported as claiming that a man believed to be covetous should not
be said to have given away everything he has all at once.[39] Again
in Tory propaganda, Churchill's personal bravery was contrasted
unfavourably with that of his ally Prince Eugene and the 'Pseudologia' recommends that a man whose personal courage is suspect
should not be reported as having led from the front, but he can, for
example, be allowed to have thrown 'a Bottle at his Adversary's
head'[40] – the likely means by which John Bull kills his first wife.[41]
Carstens may also be right to suggest that Lord 'Honest Tom'
Wharton is the target behind the advice not to report one who is

known to despise religion as spending 'whole days in his Closet at his Devotion'; and not to claim that a great man and notorious cheat has suddenly made full restitution: 'let it suffice at first, to pay Twenty Pounds to a Friend who has lost his Note'.[42] Improving a reputation 'in extremo gradu' makes lying implausible.[43] With respect to the example we are given of a 'translatory lie' (a lie in which the achievements of one are attributed to another) Pollard suggests that, again, Churchill may be the target: . . . 'on good Occasion, a man may be rob'd of his Victory by a Person that did not Command in the Action.' General Webb's victory at Wynendel, (1708) was initially attributed to Churchill's favourite Cadogan, a secretarial error out of which Tories made capital.[44]

V

All the qualifications as to probable targets, however, are important. There is not so much as a clue in the text to give any of these generalities a certainly specific focus, and we may be failing to see the wood for the trees in trying to find the exclusive references appropriate to party affiliation. There is, to adapt Pat Rogers' useful expression 'multiple implication' in the allusions.[45] The Author of the 'Pseudologia' warns the leaders of all parties against believing their lies, and either in the Projector's voice or the Author's, we are told that the Jacobites and even more so the Whigs have been much given to this of late.[46] Although the Tories are noticeable by their omission, the reader can hardly escape their implied inclusion in the phrase 'all parties'. In what, however, seems certainly to be a disingenuous allusion to the debilitating ailments of Queen Anne it is advised that if it is 'spread about that a great Person were dying of some Disease' it is more effective to lie by saying they are recovering than simply to tell the truth and say they have never suffered from it.[47] It was the Whigs who were apt to make an issue of the Queen's health. Yet if the allusion is to the Queen, what is insinuated to be the truth, namely that she is not suffering from illness, was generally untrue and none knew it better than Arbuthnot.[48] In this lie by suggestion Arbuthnot shows his Tory leanings as much as by any sly dig that might catch the conscience of a Whig and by implicating Tories achieve a roughish balance.

The Author's mooted scheme for the Whig party to regain credibility by telling the truth for three months, to build sufficient capital for further lying, is cautiously criticised by the Projector as seeming to be 'somewhat chimerical'.[49] The difference between the voices creates an illusion of considered judgement, and by this consumate economy the reader is encouraged to confront the usual Tory accusation of endemic Whig dishonesty, but there is no attempt to pre-empt or foreclose on the reader's reaction. What we have throughout, claims Beattie is an adroit and artful 'show of even-handedness'.[50] But again one must add that the show was sufficiently well sustained for it to point persistently to wider application. It may be a measure of Arbuthnot's more genuine, or thorough distance from party commitment that in the 1727 edition, this passage is significantly altered. The scheme no longer applies specifically to the Whig but to 'any party'. Re-enforcing this enhanced even-handedness, a passage in which it had been claimed in the 1712 edition that, the Tories are better believed but the Whigs have the better liars is deleted. It is replaced by the following balanced assessment: 'He (the Author) owns, that sometimes the one party, and sometimes the other is better believed; but they have both very good geniuses among them. He attributes the ill success of either party to their glutting the market . . .'[51] With too many worms, you won't catch gudgeon (see also below ch. 3, VIII).

Yet not withstanding such hooks into the topics and personalities of the day, in the context of the liberated and bracing atmosphere of party accusation, Arbuthnot, even in the 1712 edition achieves distance by the control of his irony. He has taken a step back from the polemical imperatives of peace party commitments that urge on the story of *John Bull* and has gone further down the road to what Carstens sees as a genuinely dispassionate (that is party free) satire of his later years.[52] He remains a Tory in *The Art of Political Lying*, to be sure, but a of muted and potentially subversive sort. The expression of views congenial to the bulk of Tories, at most passing jibes at demonized Whig identities, is a means to more interesting ends. There is a sufficiently well-sustained aura of even-handedness to suggest that Arbuthnot was using party labels and prejudices in part to appeal across divisions, in order to raise issues common to any group and *ipso facto* to the nature of the polity, as Swift had done more directly in the *Sentiments*, as Harley had tried, to his cost, in *Faults on Both Sides*. There is, perhaps, a symbolic intimation of the satire's range at the very beginning. The Projector lists the main coffee houses at which the work will be found.[53] All these are genuine and geographically are spread across London, a symbol

perhaps of the kingdom, from St James's to the Exchange – a
physical scope to which *The Art of Political Lying* alludes in dealing
with the dissemination of lies. Above all, however, the typical
clientele of each named club makes up a pretty fair conspectus of
political, literary and intellectual life. *St James's* had been frequented by Swift, and being very much the club of Addison and
Steele was a Whig establishment. *Young Man's* seems to have
pretended to social exclusivity, *Brydges*, or *Bridges* was a club for
commercial interests, while the *Grecean* was a club used by figures
such as Newton, Sir Hans Sloane and Edmund Halley, all associates
of Arbuthnot and was probably one of his houses as well.[54] The
reference to the work being available at any other good coffee
house across the town would encompass any group that might
otherwise seem excluded. Whether the list of coffee houses provides
an intentional clue for the reader or not, it is certainly symbolically
fitting for the potential diversity of audience, and symptomatic of
the clubbishly delineated, even divided society in which Arbuthnot
lived. The question is, how might Arbuthnot reach across so many
socially institutionalized distinctions and almost calibrated differences? How might the 'Pseudologia' make its point effectively from St
James to the Exchange, from Young Man's to the Grecean? How
might the damage Swift and before him Toland had seen group
interest and allegiance doing even to civil and ecclesiastical liberty
be counteracted? Those questions, whether or not they were exaggerated by our perspectives on the eighteenth century, take us
directly to the problem of satire in political society as I believe, it
was perceived at the time. Ironically, if the early eighteenth century
saw the emergence of a Habermasian public sphere Arbuthnot, like
the rest of the printing elite did not know much about it – quite the
contrary. Yet the way he chose to satirize his world can help create
the impression it existed.

NOTES

1   Jonathan Swift, *The Sentiments of a Church of England Man,* (1708,
     1711) *Miscellanies in Prose and Verse*, compiled Swift and Alexander
     Pope, (1727), vol. 1, p. 89; Louis Bredvold, 'The Gloom of the Tory
     Satirists', in *Pope and His Contemporaries*, ed. J. Clifford and Louis

Landa (New York: Oxford University Press, 1949), p. 7; Geoffrey Holmes, *British Politics in the Age of Anne*, (London: Macmillan, 1967), pp. 46–7 for the reluctant recognition of party conflict. Other evidence Holmes cites to show the reality of this conflict also indicates contemporaries' unwillingness to identify themselves clearly in party terms, p. 44.

2   Jonathan Scott, 'Restoration Process: Or If this Isn't a Party We're not Having a Good Time', *Albion*, 25 (1993), pp. 267ff for some salutary warnings about how much to read into the early modern notion of party.

3   Henry Horwitz, 'Party in a Civic Context: London From the Exclusion Crisis to the Fall of Walpole', in Clive Jones ed., *Britain in the First Age of Party 1680–1750*, (London: Hambledon Press, 1987), p. 173ff.

4   Holmes, *British Politics*, p. 33ff.

5   Similarly, alternatives such as Sir Lewis Namier's 'ins' and 'outs', designating men as divided mainly over the spoils of office project an altogether cooler pattern of behaviour onto a world in which the spectre of dying or killing for God still stalked. For the use of such labels in the context of Arbuthnot's work see Alexander Ross, 'The Correspondence of John Arbuthnot', Cambridge University, unpublished PhD thesis (1956), Introduction.

6   John Toland, *The Art of Governing by Partys*, (1701) p. 7ff; Robert Harley, *Faults on Both Sides*, (1710), p. 6ff; for an argument for the legitimacy of party at this time see Caroline Robbins, 'Discordant Parties: A Study of the Acceptance of Party by Englishmen', *The Political Science Quarterly*, 73, (1958) pp. 505–29. The argument makes most sense in the period following Queen Anne's reign, and Robbins wrongly attributes such an acceptance to Defoe, see Manuel Schonhorn, 'Defoe, Political Parties and the Monarch', *Studies in Eighteenth-Century Culture*, 15 (1986) p. 187ff.

7   Pat Rogers, *Literature and Popular Culture in Eighteenth-Century England*, (Sussex: Harvester, 1985), pp. 18–19; J. A. Downie, *Robert Harley and the Press: Propaganda and Public Opinion in the Age of Swift and Defoe*, (Cambridge: University Press, 1979), Introduction, pointing out that the Triennial Act must be seen in conjunction with the lapsing of the Licencing Act, p. 1.

8   Jurgen Habermas, *The Structural Transformation of the Public Sphere*, trans., Thomas Burger, assisted by Frederick Lawence, (London: Polity Press, 1992), ch. 3.

9   Rogers, *Literature and Popular Culture*, pp. 19–22; Holmes, *British Politics*, p. 30 refers to 18 private newspapers in London by 1709 in addition to party and government sponsored sheets. See also Downie, *Robert Harley and the Press*, pp. 10–15.

10   J. A. Downie, 'The Development of the Political Press', in Clive Jones ed., *Britain in the First Age of Party, 1680–1750*, (London: Hambledon Press, 1987), pp. 111ff esp. 125–6.

11   Charles Davenant, *Tom Double Against Dr D-v-n-t*, (1704), p. 13; Downie, *Robert Harley and the Press*, p. 8 on the tendency to create

propaganda images of extremity; J. G. A. Pocock, *The Machiavellian Moment*, (Princeton; University Press, 1975), ch.13.

12   B. Lillywhite, *London Coffee Houses*, (London: Allen and Unwin, 1963) provides a detailed study. It is clear that the social functions of coffee houses were diverse, they played a role in the development of insurance, the post and banking; and that individual houses could be highly specialized. Some were very clubbish and although Judge Jeffreys had wanted those circulating dubious newspapers closed in 1688, it is difficult to generalize as to their overall role in the circulation of information through society, see pp. 17–21; also Sir George Clark, *The Later Stuarts, 1660–1714*, (Oxford: Clarendon Press, 1961 edn) pp. 358–9.

13   Holmes, *British Politics*, p. 30; Downie, *Robert Harley and the Press*, p. 1ff; 'Development of the Political Press', p. 115.

14   Isaac Kramnink, *Bolingbroke and his Circle*, (Camb. Mass.: Harvard University Press, 1968), p. 10; Carstens, 'Political Satire', p. 17ff; for an extensive discussion, J. A. Downie, *Robert Harley and the Press*.

15   Holmes, *British Politics*, p. 75–81. Downie, however, notes that the Sacheverell affair of 1710 generated some 500 separate titles, which would indicate that the wars had some competition; Downie, *Robert Harley and the Press*, p. 10. Toland, *An Appeal to Honest People Against Wicked Priests* (1710) held that priests like Sacheverell would 'settle Non resistence by Rebellion' p. 31; they had always been the causes of civil wars, p. 56.

16   See at length, Downie, *Robert Harley and the Press*, e.g. p. 135; D. Szechi, *Jacobitism and Tory Politics, 1710–14*, (Edinburgh: Donald, 1984) ch. 5; J. R. Jones, *Court and Country: England 1658–1714*, (London: Arnold, 1978), pp. 343–4, citing Harley. See also Robert Harley, *Faults on Both sides*, p. 4; and the reiteration in the *Vindication of Faults on Both Sides*, (1710), p. 27.

17   Anon., *Most Faults on One Side*, (1710); *Faults in the Fault-Finder* (1710), followed by a *Supplement* (1711). See also Downie, 'Development of the Political Press' pp. 115–6; also more generally, Holmes, *British Politics* pp. 48–9.

18   Downie, *Robert Harley and the Press*, p. 135 suggests that during 1711 Swift was siding more with the hardline Bolingbroke.

19   Swift, *Sentiments*, p. 91; see also Toland, *The Art of Governing by Partys* (1701), p. 58.

20   Swift, *Ibid.*, p. 94.

21   Toland, *The Art of Governing by Partys*, saw party division as a sort of civil war, a figurative hyperbole that made it easier to conclude that the consequence of party government was the development of arbitrary power out of the exhaustion, pp. 60–1, p. 7.

22   Swift, *Sentiments, pp. 113, 114; cf Harley, Vindication*, p. 3 on bigotry and adherence to party.

23   Davenant, *Tom Double*, pp. 3, 4.

24   Swift, *Sentiments, p. 137*.

25   Holmes, *British Politics*, p. 30.

26 Davenant, *Tom Double*, p. 21; cf Swift, *The Public Spirit of the Whigs* (1714) 'Whigs cry out 'A Pamphlet! A Pamphlet! The Crisis The Crisis, not with a view of convincing their Adversaries, but to raise the Spirits of their Friends . . . and unite their Numbers . . . as Bees assemble and cling together by the Noise of Brass' p. 6.

27 Patricia Carstens, 'Political Satire in the work of John Arbuthnot', London University unpublished PhD thesis. (1958), p. 12; Downie, *Robert Harley and the Press*, p. 1–2, 10; Holmes, *British Politics*, pp. 75ff.

28 Szechi, *Jacobitism and Tory Politics*, ch. 5. for the shifting complexity of Harley's Tory support.

29 Arbuthnot, *A Sermon Preach'd to the People at The Mercat Cross of Edinburgh; On the Subject of Union*, (1706); Daniel Defoe, *An Essay at Removing National Prejudices Against a Union with Scotland*, (1706). I am indebted to Bridget McPhail's unpublished paper, 'Daniel Defoe and the Construction of 'Great Britain', (1994).

30 Szechi, *Jacobitism and Tory Politics*, e.g. pp. 109–110, 194; Downie, *Robert Harley and the Press*, at length; but see also Jones, *Court and Country*, p. 346–7, for his peremptory treatment of court Whig enemies.

31 See above notes 16, 17.

32 Cited in Lester M. Beattie, *John Arbuthnot Mathematician and Satirist*, (Camb. Mass.: Harvard University Press, 1935 and Russell & Russell, 1867), p. 20.

33 Arbuthnot, *John Bull, Appendix to John Bull Still in his Senses*, ch. 3, p. 87; Beattie stresses correctly that on the evidence of *John Bull*, Arbuthnot was not at one with mainstream Tory views, but the Harleian association is undeveloped. See *John Arbuthnot*, p. 160; Bruneteau, 'John Arbuthnot (1667–1735) et les idées an debut du dix huitieme siècle,' Doctoral thesis Université de Lille (1974) vol. 2, pt. 4, ch.2 p. 687 does note the pamphlet's affinity with Harley.

34 Carstens, 'Political Satire', p. 301.

35 Carstens, *Ibid.*, pp. 296, 318; despite the reference to Harley, Bruneteau follows, 'John Arbuthnot et les idées', pt. 4, ch. 2.; pt. 5, ch. 1 pp. 723–4.

36 Carstens, 'Political Satire', pp. 298ff. The most extreme statement is that '(a)s all the lies discussed in "the Fifth Chapter", with the exception of the remark about the French king, have been whiggish, it seems that the Author directs his remarks to "all Practitioners" under the assumption that they are mostly Whigs!' p. 308. There is no justification for glee at this sort of logic. It contradicts what the Author says about the People and their rights, about parties generally, about the necessity to combat lies with lies. It also sits ill with the subtleties of much of her analysis.

37 Davenant, *A True Picture of a Modern Whig, Set Forth in A Dialogue Between Mr. Whiglove and Mr Double*, (1701) p. 9.

38 Arbuthnot, *APL*, pp. 14–15; Carstens, 'Political Satire', pp. 309–310, cf. Davenant, *The True Picture of a Modern Whig*, 1701, p. 9 where he writes of the over-use of these lies.

39 Arbuthnot, *APL*, pp. 12–13.

40 Arbuthnot, *Ibid.*, p. 13; Carstens, 'Political Satire', p. 302.
41 Arbuthnot, *Law is a Bottomless Pit*, ch. 8, p. 14.
42 Arbuthnot, *APL*, p. 13; Carstens, 'Political Satire', p. 303.
43 Arbuthnot, *APL*, p. 12.
44 A. F. Pollard, *Political Pamphlets*, (London: Routledge, Kegan, Paul, 1897), p. 112; George Aitken, *Life and Works of John Arbuthnot* (Oxford: Clarendon Press 1892), p. 297.
45 Rogers, *Literature and Popular Culture*, p. 12.
46 Arbuthnot, *APL*, p. 19.
47 Arbuthnot, *Ibid.*, p. 21; Bruneteau, 'John Arbuthnot et les idées', pt. 1. ch. 2. p. 70 cites Peter Wentworth on Arbuthnot's desires to keep the Queen's illnesses secret. See also Ross, 'Correspondence,' letter 35 (Sloane 4043 fol 76) vol. I, p. 225 on the good health of the Queen.
48 Even the qualification 'generally' is important, Harley was not above encouraging the Queen to an attack of diplomatic gout to slow down procedures, Szechi, *Jacobitism and Tory Politics*, pp. 109–10.
49 Arbuthnot, *APL* p. 16.; Carstens, 'Political Satire', p. 303.
50 Beattie, *John Arbuthnot*, p. 293.
51 See for example, Pollard, *Political Pamphlets*, pp. 116–117; Bruneteau, John Arbuthnot et les idées,' pt. 5. ch. 1 pp. 723–4; Carstens notes the change, 'Political Satire', p. 313 but misses the point which in general terms she insists upon.
52 Carstens, 'Political Satire', p. 327.
53 *APL*, p. 5
54 Lilleywhite, *London Coffee-Houses*, pp. 500–1; 668, 223–4; 172–3, provides the details about the clubs on which I have drawn; (see also below, Appendix C, n. 1.). According to Downie, *Robert Harley and the Press*, the Grecean had been a point of distribution for Harleian propaganda, p. 36.

3 Satire

Arbuthnot made me draw up a sham subscription for a book, called a History of the Maids of Honour since Harry the Eighth, showing they make the best wives, with a list of all the Maids of Honour since &c to pay a crown in hand, and t'other crown upon delivery of the book; all in the common form of those things. We got a gentleman to write it up for us, because my hand is known, and we sent it to the maids of honour when they came to supper. If they bite at it, 'twill be a very good court jest; and the queen will certainly have it; we did not tell Mrs Hill.'[1]

Jonathan Swift, The Journal to Stella, 19, Sept., 1711.

I

A few days later, after relating this prank and after the maids had clearly bitten, and the story was out, Swift recorded 'That rogue Arbuthnot puts it all upon me'.[2] The following year on February 19th Addison announced a new French Academy for the education of politicians in all facets of their dissimulatory craft, in gestures and diplomacy and the whole language of lying, 'French Truth'(see below, ch. 7, VI).[3] Later, *Plain Dealer* outlined the contents of a proposed six volume history of recent Whig lies.[4] The first volume was to contain 'A full Account of the Nature, Property, and Usefulness of Lying'. The second concerned the art of mastering and perfecting 'all Sorts of Lyes, of what Size or Quality soever'. The third dealt *inter alia* with improbable lies, scandal, treasonable insinuation, city rumour and dogmatic assertion. The last three volumes comprised a history of the lies of the last few months 'Collected from the Mouth of the Whiggish Party, viz. Papists, Republicans, Atheists, Deists, Socinians, Independents, Quakers, Anabaptists, Sweet-singers, Muggletonians, French-Prophets, and a thousand other different Sectaries. With a large Index to the whole.' As Beattie remarks, this announcement was a probable 'spring-board' for Arbuthnot's own 'Curious Discourse'.[5] It is, of course, considerably more blunt than Arbuthnot was to be. Yet because of this, the bogus advertisement reveals more obviously something of the slippery semantics surrounding accusations of lying, (see below ch. 7, IV, V) and further evidences the importance of the rich and divisive vocabulary of party hostility.

Shortly after this trumpeting Tory advertisement, Pope printed a
proposal for the serial publication of 'The works of the Unlearned'[6]
Given Arbuthnot's continuing interest in science and the consolida-
tion of knowledge, this too may have provided a springboard. These
parodies were parasitic upon a buoyant advertising trade, especially
in books.[7] Swift passingly noted the similarity to 'those pamphlets
they call the works of the learned.'[8] Patricia Carstens notes a genuine
article particularly relevant for the form of *The Art of Political
Lying*'s own enticement to buy: 'Proposals for printing the Memoirs
of Philip de Comines:.... Is now in the Press, and almost finish'd
in order to deliver to the Subscribers at Michaelmas next. All
Gentlemen that are willing to encourage so excellent a Work, are
desired to send in their Subscriptions .... Subscriptions taken in by
J. Phillips, at the Black Bull.... The whole work consists of 75
sheets, in 2 vol. 8vo at 10s in Quires, 5s down, and 5s on Delivery.
Those that subscribe for 6, shall have a 7th gratis.'[9]

I shall only note the ironies of responsibility in some of these
points of resemblance which seem to anticipate the entangled
authorships of Swift, Arbuthnot and Pope. What Swift called the
common 'form of those things', the subscription advertisements of
the book trade, is carried through from the private court jest to
become the indispensable means by which Arbuthnot structured his
satire. From the 'History of the Maids' to the 'Pseudologia', (via
the 'History' of Whig lies and 'The Works of the Unlearned' the
price has gone up. Seven Shillings, the advertisement informs us, is
to be paid down, seven on delivery, with the backhanded promise
of printing the full details and addresses of those who subscribed.[10]
The 'form' of those things, is in this case a perfectly genuine fake.
There are none of the exaggerations or other clues that indicate a
joke or signal a parody as with 'The Works of the Unlearned' or
the proposed six volume history of Whig lies since last winter.
When *The Art of Political Lying* was itself advertised, ironically
planted in the Whig *Daily Courant*, the solemnity was maintained.
Under the heading 'Just Published' was announced the title 'Proposals for Printing... with an Abstract of the Volume of the said
Treatise.' The printer's name, location and the price were stated as
was the norm.[11] And there is an added joke, in that Swift, in
sending the manuscript to be printed (price 3d) was helping to
create a simulacrum of something he was to state that he thoroughly disliked (at least when produced by Bishop Burnet), namely
cheap pamphlets acting as promotional outriders to hefty tomes.[12]

II

In order to sustain the semblance of truth to the initial claim that there was a book in press, Arbuthnot needed to remain indirect. Formally there are two senses of indirection relevant to *The Art of Political Lying*. There are devices which help distance an author from the satiric persona or target in the text; and there are forms of satire in which the force of the satire itself is indirect or ambivalently focused. In practice these forms of indirection are likely to be mutually entangled. I shall say only enough about each to cast light on Arbuthnot's satire.

Authorial indirection itself had a number of manifestations, from simple anonymity, translation and pretended translation, to history, allegory, the fiction of false authors within or authenticating the principal text, and the imitation of previous satirists, this last a process which at once established satiric traditions of discourse and drew cutting parallels with earlier times.[13] Although one motivation, or pretended motivation lay in the fear that the satiric prophet would be persecuted or prosecuted, which itself makes a double rhetorical point,[14] authorial indirection had other benefits and by the early eighteenth century amounted to a self-sustaining literary habit. One advantage, as Carstens points out, was to broaden the scope of what might otherwise be only narrowly topical.[15] This, however, might well result in indirection of satiric force. As Carstens also remarks, Arbuthnot was particularly at ease with modes of indirection by which all of his satires are marked.[16] In one consistently sustained fashion, *John Bull* is authorially indirect simply because it is a satiric allegory; the specific targets of the satire always require decoding from the story which proceeds independently of them.[17] Thus, when he is satirizing theories of political revolution, the vehicle is a speech on marriage; when political discourse is found directly in the text, it is there as a vehicle for a satire on Calvinist theology. The result of this, together with the persistent tone of intimate good humour, and the well rounded character of John Bull himself, enables Arbuthnot, to some extent, to appeal across party fissures and display a sophisticated awareness of the diversity of his audience.[18] But the indirection created by a fake authorial persona is not consistent. The character of Sir Humphrey Polesworth begins to assume an importance, helping to explain the style and apparent imperfections of the work only from *John Bull Still in His senses*. Despite these devices of authorial indirection, the satiric force of the pamphlets is, generally clear and pointed (for qualification see below ch. 6) The war party is wrong, immoral and corrupt; peace is good and in England's interest. The Scots may be tiresome, but they have been treated shabbily in the past, a similar peace to consummate the Union is in everyone's interest.

III

By contrast, in *The Art of Political Lying*, there is no allegory, satiric or otherwise, the work is directly about the nature of political discourse. There is, however, a double authorial indirection created by the pretence of there being a book which a third party is promoting. Arbuthnot was to use this device again in *A Brief Account of Mr. John Ginglicutt's Treatise*, (1731) but in a simplified fashion. The *Brief Account* which satirizes the decline of the discourse of abuse, is within the frame of a proposed subscription unevenly described by the 'author' himself.[19] In *The Art of Political Lying* the indirection necessitated by the initial advertisement is fully sustained and requires, what I have already noted, the dual textual personae of the satire in addition to the anonymous Arbuthnot himself as creator of the whole. Through the persona of the Projector, Arbuthnot achieves the aura of distance; the Projector admiringly recounts to the audience of potential dupes what the Author is apparently arguing with what Bruneteau aptly styles 'an imperturbable gravity'.[20] This creates something of the multiple-voiced potential of the dialogue with a few of those discrepancies one might expect from what Carstens refers to as an alliance between Pedantry and Grub Street.[21] Who is fooling or pretending to fool whom remains in tension, for Arbuthnot never lets down his guard. And indeed the work ends on a totally appropriate note of anticlimax, with the bland statement that '*The Account of the Second Volume of this Excellent Treatise, is reserv'd for another time.*' The text itself is parenthesized between falsehoods expressed as a hope and a promise. Between the hope of a successful subscription (followed immediately with a disingenuous assurance of delivery) and the promise of an account of the second volume, Arbuthnot sustains a satire which is impersonal in two senses. It is easy for a learned treatise to be presented as operating at a level well above personality and reproach, as general satire (see below section V). It is after all a scientific treatise.[22] It is also possible for Arbuthnot to create and sustain an air of disinterestedness and balance both with respect to the political world and the theories being presented about it by keeping the Projector between himself and the Author.

Additionally, however, because we have only a tantalizing sketch of the contents of the 'Pseudologia', Arbuthnot need only suggest details, so flattering the understanding of the reader and offering an invitation to fill lacunae. 'The Tenth Chapter treats of the Characteristicks of Lyes . . . Your *Dutch, English* and *French* Ware are amply distinguished from one another.'[23] With no more detail, we can only imagine what the differentiae of *our* Dutch lie might be. Indeed, we are told, the Author wishes to be informed of any new discoveries and invites the reader to help in the further perfection of the theory.[24] For almost all of this, however, we have no more than the Projector's word. Only once does he 'quote' from the original. He twice demurs at the Author's argument, which evidences independence of judgement, but his claim to have read the work with great care is perhaps undercut by his referring to two chapter eights.[25] In this way, through a non-confrontational approach and encoded invitations to scepticism, Arbuthnot can appeal across Swift's friar-like affiliations and inveigle the reader into a tacit and critical participation as the account unfolds.[26] If perchance the fiction is swallowed, the presupposition that politics is a fundamentally dishonest business is taken down as well. If we merely appreciate the joke, say as Tory sympathizers, we are nevertheless nudged towards a critical perspective on the political world of which we are a part. To recognize these potentialities, is to see the adaptation of a theory of persuasion to political satire, and to begin to isolate the indirection of force which is so crucial to the work. The rationale for this mode of indirection, however, is entangled with the crisis of satiric success that was so evident in the early eighteenth century.

IV

In an image closely akin to the optical ones favoured by Arbuthnot, Swift remarked that satire was popular precisely because it was like a glass in which we could see everyone but ourselves.[27] This, in a sense, was a sign of failure rather than success – satire was a

popular success because it was a moral failure.[28] As Samuel Butler
had earlier sought to expose a party type in *Hudibras* (1663)
without real hope of bringing about reform, so Swift generalized
about the problems of satiric correction which were worsened in an
age of party.[29] In this way, Swift's decorous visual simile of a state
of moral purblindness was also exactly consistent with a main
thrust of his *Sentiments*, that men were currently ill-equipped for
rational and critical judgement. Being embroiled in party allegiances, they could but mindlessly attack each other (see above ch. 2,
II). If, therefore, satire allows us all to see human folly and
corruption everywhere but in ourselves, how can it reform? To be
sure, it might plausibly hope for some impact by altering the
reputation of a potential victim in the mind of the reader and this
would be no mean feat in an age in which social reputation could
approach being social identity (see below, ch. 7, V). Nevertheless,
the genuine moral reform in which satirists had such a heavy
legitimising investment required much more in a world of imper-
vious sinners, it required a reflective and critical sense of self as well
as a sense of self through the other. In short and in general, unless
satire promotes the capacity to know oneself, it can do little more
than reinforce the pride of the morally censorious throughout
society, degenerating into an organ of party propaganda. As Hobbes had put it, if you would understand others, 'Nosce teipsum,
Read thy self'.[30] However satirists might castigate such failings as
hypocrisy and pride, they write from a sense of pride, a confidence
that they occupy a high moral ground from whence they work at
best to reform or at least censure lesser mortals. Truth, as they have
almost ritually affirmed since antiquity, is fundamental to them and
it is the confidence that they are in possession of a sufficiency of it
that enables them to proceed. The consequent rationale for satire
had been that, however amusing and or destructive it might seem,
it was designed not to wound or destroy for its own sake but for
some greater good.[31] From pride there came a moral obligation to
improve others and, in so far as there was a genuine moral outrage
at the times, and not a mere hypocritical satiric mask for party
and personal vindictiveness, men such as Swift, Pope, Arbuthnot
and before them Dryden and Hobbes, were bound to ask how
satire could be effective. The ineffective satirist was too easily
seen as trivial and stained by the very presumption he characteristically attacked. Just so Dr Johnson rebuked the whole Swift
circle for thinking they had a monopoly on virtue.[32] As any one

who wrote about it knew, satire was a morally, as well as potentially a politically dangerous game. Its ambivalent place in Sidney's hierarchy of literary form was always precarious, it was thus in periodic need of reassurance, rationalization and promotion.[33] The satire of Augustan England was no exception, indeed, that age of satire generated a great deal of hostility to satirists. Often elided with libellers and common detractors writing from the lowest of motives, they were above all held, as Peter Elkin has shown, to be counterproductive a menace to themselves and society at large.[34]

What was required of satire and what means it might employ for its ends, had for a long while been animadverted upon. The variety of satiric style from the Renaissance in England alone was striking.[35] The result had been a very open-ended phenomenon, sometimes little more than a dimension in more easily discernible forms of writing.[36] Nevertheless, during the sixteenth and seventeenth centuries satire had been predominantly scathing, its force direct; but as satirists sometimes suspected, rarely effective at reforming conduct.[37] Satirists might seem to agree in the most abstract terms that we live in a naughty world and vice is to be reprimanded, what divided them and extended the very notion of satire was how this was to be done.[38] Could satire embrace invective and abuse, humour and exhortation? This degree of specificity as to means, de-stabilized easy assumptions about ends and rubbed at the very edges of a clear satiric identity. Satire could be found in anything from slander to the sermon. In the case of Thomas Hobbes, it is found in formal philosophy, as it was to be again in *The Memoirs of Martinus Scriblerius*.[39]

V

With respect to tone, satirists could be scathing or relatively gentle associating themselves respectively with the authority of Juvenal or Horace. This of course could be a matter of degree and variation within a single work. Elkin has argued that the eighteenth century saw an overall shift towards the urbanity of the Horatian. If there was such a shift, it is best seen as a change of tactical emphasis to aid reform in a fissiparous world.[40] In so far as there was not, a Juvenalian tone is largely at one with the heated atmosphere of debate in which Arbuthnot wrote. Either way, Arbuthnot's satire is persistently Horatian, indeed so subtle and urbane that it slips through the course tangle of modern definitional nets.[41]

Another now familiar distinction had emerged, which divided Arbuthnot especially from Pope. Superficially, this involved opting for personal, what Dryden called private satire, or for general satire, a choice as to both means and ends.[42] Rarely is there any clear-cut dichotomy involved but rather locations of emphasis along a satiric continuum.[43] The general may be styled satiric realism in as much as general human failings, pride, hypocrisy, mendacity and so on are the prime objects of attack. Confronting them, directly or indirectly, through a Juvenalian harshness or an Horatian urbanity, was the appropriate means to satiric success. And satiric realism had, albeit from motivations of caution, proved successful for Erasmus and Sir Thomas More.[44] Joseph Hall noted the virtues of generality which, hand in hand with a sensitivity to audience, invite us to consider if any strictures apply to us. 'Art thou guilty? Complain not, thou art not wronged. *Art thou guiltless?* Complain not, thou art not touched'.[45] Robert Burton, alluding to Erasmus put the matter with rare succinctness: 'I hate their vices, not their person'. Let him, he continued, be angry with himself if guilty, 'If he be not guilty, it concerns him not . . . '[46] By contrast, personal satire thus becomes a form of nominalism, what really matters is to pin-point the individual sinner.[47] Or, in the case of the nominalist philosopher Hobbes, the targets are priests, lawyers and the democratical gentlemen of parliament, that is, all the individuals within a specific classification. As with debates over the primacy of realism and nominalism, to make a case for one extreme requires presupposing the significance of the other, for they are, as it would have been put in the seventeenth century, relational terms. There is, then, a sense in which general satire without cases, reference to specific failings is at one extreme of satire, a cold skeleton of values to be fleshed out; safe, (perhaps) because of its generality, it may by this token alone be ineffective.

At the contrasting pole, personal satire, by evoking shared values, presupposes the general, but exists on the brink of diatribe, calumny and slander;[48] dangerous because of its direct force and particularity, it may be similarly ineffective as more people are able to exclude themselves from its vision. As James Bramston was to put it, 'Let pamphleteers abusive Satyr write,/To shew a Genius is to shew a spite'.[49] Additionally it might only give welcome fame to those who would otherwise be consigned to oblivion.[50] Perhaps the truth is that the distinction between general and particular was less concerned with types of satire than with issues of defence and promotion. It does, for example, seem to have been Erasmus's belief that an emphasis on the general, which thus could be focused on nothing more specific than the hypothetically populated class rather than upon the vindictive individual was more prudent in a dangerous world. Yet there was no guarantee that the offended individual would recognise the mis-fit of a critical cap.[51] This was a possibility which seems not to have escaped cautious Robert Burton, who having stressed that the innocent have nothing to fear immediately begs to be excused for any folly.[52] Similarly, there is at least an element of self-promotion in Pope's defence of personal satire. In terms, however, of any necessary nature of satire there is, as with any complex character, no absolute choice between a variable balance of dispositions.

VI

It became Arbuthnot's view at some time that general satire was to be preferred to the more personal. In this respect the balance shifts between *John Bull* and the *Art of Political Lying*. In the former the display of villainy through an allegorical narrative presents a Drydenesque way of excoriating particular sinners. In the latter, the generality helps sustain the ethos of disinterest, and the discretion of the illustration that stops short of pinning the hapless victim to the page, helps create an indirection of force. *The Art of Political Lying* is sufficiently close to the skeletal extreme of generality to require the reader to flesh it out, and as it were, breathe satiric life into it by imaginative application, to re-direct the force. In contrast, Swift's attack on Wharton (*Examiner*), sharing both the publisher and the *topos* of lying is close to the personal pole.[53] Framed in general reflections on 'the Art of *Political Lying*'[54] of the sort that Arbuthnot was to elaborate, the focus is so centrally on one great un-named person, that the piece might safely be read as about the recently defunct lord. 'The Superiority of his Genius consists in nothing else but an inexhaustible Fund of *Political Lyes*, which he plentifully distributes every Minute he speaks . . . . He never yet considered whether any Proposition were True or False, but whether it were convenient for the present Minute . . . . The only Remedy is to suppose that you have heard some inarticulate

Sounds, without any Meaning at all.'[55] Only once after *The Art of Political Lying* was Arbuthnot to give himself over to a full blown personal attack on another human being. His *Epitaph on Francis Chareris* was a lone excursion into the Juvenalian and even then, as the victim was dead and beyond reform, he assumed the status of a type.[56] The *Quidnuncki's*, an allegorical beast fable usually attributed to Arbuthnot, but possibly by John Gay, sustains the form of a fable recounting a society of monkeys climbing a tree until one getting to the top falls into the water to the temporary consternation of the rest and so on *ad infinitum*. It is a withdrawn Augustinian reflection on the exaggerated importance of politics in general and is appropriate to the views of either Arbuthnot or Gay.[57] *John Ginglicutt's Treatise* though probably provoked by recent excessive name-calling which had stimulated a duel of honour between erstwhile friends Hervey (shortly to become Pope's 'Sporus') and Pulteney, gets its life from marshalling the flytings of ancient orators who were not so foolish as to let the conventions of insult run away with them.[58] It is Mandevillian and Hobbesian in its distaste for what Arbuthnot styles in Harringtonian fashion, the 'gothic' politics of honour, but he could hardly make his point about insults if his satire became personal.[59] As Beattie sums up, Arbuthnot lacked the aggressive and socially driven satiric impulse of Swift or Pope.[60] The general character of politics, the disintegration of a public activity are the themes Arbuthnot maintains from *The Art of Political Lying* on and as I shall argue, satiric generality is reinforced by the alterations to that work's re-printing in 1727.

Thus, when Arbuthnot came to co-operate with Pope on *Peri Bathous, or the Art of Sinking in Poetry*, he wished to attack the literary sins of the age not the sinners. *Peri Bathous* is most safely seen as a piece of Scriblerian satire. Written by the Scriblerian Club's eponymous hero Martinus Scriblerus, it satirizes, false and misdirected learning and the naive mechanical reductionism the Scriblerians associated with bogus science.[61] It was probably initiated in the early stages of Scriblerian activity and then taken up again by Arbuthnot and Pope.[62] Arbuthnot apparently worked on it while Pope was writing the *Dunciad*, but despite Pope's blandishments, did not finish it and it was Pope who polished and revised it and found it very good, especially as a satire on the *ars rhetorica*.[63] This alerts us to the fact that it has more than just a title that echoes *The Art of Political Lying*.[64] As Kerby-Miller and Bruneteau have suggested, *The Art of Political Lying* may form part of its literary context.[65] It is enough now simply to note that *Peri Bathous* has a particularly strong family resemblance to *The Art of Political Lying* (see Appendix B). The main point here, however, is to emphasize a difference in satiric style between Arbuthnot and Pope despite the structural similarities between *The Art of Political Lying* and the fruits of their Scriblerian collaboration. The Scriblerians were bound more by the perceived objects of their satire than by a shared theory of satire as such or by a unifying tone. Pope practised a highly personalized nominalistic satire and believed that general satire was ineffective. Such a satiric idiom had already made for troubled collaboration within the Scriblerian circle. *Three Hours After Marriage*, predominantly by Gay but with assistance from Pope and Arbuthnot, was met by a probably orchestrated chorus of outrage because of its personal satire. Even in advance of performance it was rumoured that Pope was about to attack every writer of the age.[66] The play's relative lack of success after a comparatively long if controversial first run, was blamed on its somewhat exaggerated propensities to pointedness and particularity, with Pope more than Gay and far more than Arbuthnot being held responsible.[67] Pope remained undaunted and in the year after the publication of *Peri Bathous* he produced his immediately popular and more scathing *Dunciad*, to which Arbuthnot had nevertheless contributed notes.[68] Authorial identity and satiric emphasis continued to overlap and so, especially on the basis of such incomplete evidence, one should resist any notion of a neat trajectory of change or any division into rigidly opposed satiric parties. Yet Arbuthnot may have procrastinated in the joint project because of memories of the bruising if bracing experience of theatrical involvement with Pope and because of intermittent differences of satiric strategy. Satiric differences between the two were hypothesized at the time;[69] and shortly before his death Arbuthnot wrote to Pope requesting in passing that while Pope continue to abhor vice, he place the emphasis of his satire on reform more than chastisement, difficult though it was, he accepted, to separate the two.[70] In an image revealing a typical Juvenalian confidence, Pope replied that the animals had to be cut from the herd. 'General satire in a time of general vice has no force'.[71] To recall Swift, one may paraphrase: if in an age of corruption people do not recognize themselves in the glass, its light must be made to shine mercilessly upon them. In a more considered reply designed for public consumption and self-defence, Pope revealed his awareness of the conceptual

interdependence of general and personal satire and his own in-
eradicable, courageous bias in favour of the latter, and all in a way
that collapses any unstable distinction between satiric means and
ends. 'But sure it is as impossible to have a just abhorrence of Vice,
without hating the Vicious, as to bear a true love for Virtue, without
loving the Good . . . . To attack Vices in the abstract, without
touching Persons, may be safe fighting indeed, but it is fighting
with Shadows. General propositions are obscure, misty, and uncertain . . . . Examples are pictures, and strike the Senses . . . raise the
Passions, . . . to the aid of reformation.'[72]

General satire, in other words, has too dissipated a force to be
effective. The *Epistle to Dr Arbuthnot*, is the most formal statement
of Pope's opinions on the inadequacies of general satire, albeit
presented with an ameliorating nod in the direction of Arbuthnot's
views which helps give the work a less persistently Juvenalian tone
than the letters might lead us to expect. It had been long sketched
in but was forced into completion by his old friend's impending
death, appearing a month before Arbuthnot's demise.[73] In this it
remains true that Pope recognized some interdependence of general
and particular, and the attack on Sporus was so vicious because of
Pope's abhorrence of all that he stood for (line 325); but the
emphasis on the courageous and purgative particular was maintained, with the Erasmian voice of Arbuthnot accentuating Pope's
argument by worrying about the dangers of being offensive (lines
75–8). Yet, in this poem, as Knopflmacher has pointed out, Arbuthnot himself was reduced to a type, the physician of the body as
opposed to Pope's self-projection as a true satirist, a physician of
the soul.[74] If this hardly did justice to the historical Arbuthnot, it
also rendered the problem of satiric effectiveness too one dimen-
sional. What Pat Rogers has argued to be the poem's prevailing
tone of anxiety may in part extend to embrace the emblematic
reduction of Arbuthnot and the concomitant bravura of Pope's
self-image of satiric integrity.[75] It brings me directly to a further
distinction that helped draw the line between friends over the most
efficacious strategies of satire.

VII

Ancient rhetoric was predicated on the belief that any topic for
discussion could generate diverse reactions, that it was always

possible to argue *in utramque partem*, on both sides of a case – there could be no rationale for persuasion otherwise. From this condition of rhetorical discourse arose a more specific and contingent strategy of presenting differing views for more effective persuasion. This involved that ancient distinction originating with Zeno between the utility of the open palm of rhetoric as opposed to the closed fist of logic. The one was indirect and designed to entrap, outflank or even surprise the audience into a greater self-awareness and thus persuade; the other was designed to confront and destroy opposing arguments. One was a matter of manipulating emotions and understanding an audience, the other a matter of presenting propositions as if they were persuasively sufficient unto themselves; or more precisely, as if all audiences were equally and fully committed to the same tightly textured forms of discourse. A rhetorical theorist such as Cicero was in a sense bound to prefer the open palm because so much of his theory of rhetoric concerned the variable tactics of making sound arguments effective. Were he to have supposed a world populated only by formal logicians, *De oratore* would have been somewhat shorter than it is.

The metaphor of fist and palm remained central to the characterization of rhetoric, ' ... logic differeth from rhetoric ... as the first from the palm, the one close, the other at large ... '.[76] As satire was persuasive, the relevance to it of these opposing argumentative strategies was by no means new in the early eighteenth century. In his *Letter to Dorp*, Thomas More had discoursed at length on the argumentative choices to be made, and his opting for the satiric open palm of rhetoric to encourage reflection and overcome prejudice may be important in understanding the structure of *Utopia*.[77] Erasmus too had intimated that as Dorp understood only 'propositions, conclusions and corollaries', more subtle indirect and disarming satiric methods were required to make him think.[78] Lorenzo Valla had elaborated his own version of this in his argument that whereas philosophy proceeded in direct and linear fashion, rhetoric presented the audience with such a diversity of arguments that where one might not convince, another might, where one might fail, another could be brought up to take its place. The rhetorician is, in short, better and more fully armed. Rather than moving all the forces in a straight line, all predicably and thus vulnerably dependent upon each other, the rhetorician commanded with diversity, dexterity and surprise.[79] Nothing, as Burton was to remark, out of Scaliger, more invites the reader than the argument

unlooked for and unthought of.[80] Roman rhetoricians had shown
an acute awareness of the need to vary cases according to the
audience, not just by taking an account of the differences between
judge and jury, in forensic rhetoric, but more generally by introducing different types of argument and varying the appeal for say, the
altruistic or the self-interested.[81] Bacon, among others had reiterated
the point and found precedent enough in the New Testament; and
in so far as the audience was heterogeneous and under-determined,
this whole issue of intellectual dexterity could raise delicate
strategic problems for the rhetor, with severe problems of coherence
for the presentation of arguments in texts.[82]

For the most part Renaissance and seventeenth-century English
satire had been direct, even angry, its use of the tools of rhetoric,
appropriate to Valla's pretty commonplace combative imagery of
warfare. The integrity of the satirist might be in doubt, but not
where he stood implacably opposed to vice, this was something as
true of Donne as it was of the notoriously evasive Marvell and of
the egregious Hobbes discoursing on the overriding necessity of
peace. As Dryden instructed, there should be unity of theme in
satire and a single moral precept to guide against the follies
discussed.[83] Hobbes's notoriously arrogant persona seemingly alluded to by the opening claims of Arbuthnot's Author, (see below,
ch. 4, i). was itself typical of Renaissance satiric and eristic pride.
Hobbes may have developed the satiric dimension to the arguments
of *Leviathan* because, as Swift was to become, he was deeply
sceptical of the efficacy of rational argument, the closed fist of
logic, in a clamorous world of divisive and self-deluding factional
interests.[84] He may, indeed, be understood as experimenting with a
diversity of rhetorical techniques used to satiric and other effects in
order to overcome the limitations of rational argument common to
the differing sections of his potential audience.[85] The experimenta-
tion, this Valla- esque variation on the theme of the 'open palm' of
rhetoric was not an unalloyed success. Dismissed as a scoffer who
would not argue rationally enough, Hobbes blunted or removed
many of the satiric barbs in the Latin edition of his work.[86]
Leviathan and its immediate fate, in short, stands as early testimony
to the problems Swift saw in the attempt at satiric reform. As I will
suggest, in this way it is indirectly relevant to the answer to those
problems offered in *The Art of Political Lying.*

One tract attributed to Arbuthnot claimed that for the first time,
every form of satire existed on an equal footing,[87] perhaps as they

might have seemed to be within the compass of *Leviathan*. Notwithstanding hyperbole, this claim underscored the importance of rhetorical dexterity and a sensitivity to group prejudice and it raises
the question of whether it is simply a lack of perception that leads
us all to see only other faces in the glass. It was certainly likely that
direct attack, which as Pope had it, 'raises the Passions' simply risks
raising the defences of 'interest' as Hobbes had found, with the
clerics he so roundly satirized, dismissing him in turn. Thus satire's
intended victim would be able to redescribe or deflect the attack as
vindictive; to construe the pointed exaggeration as literally missing
the mark and so therefore being altogether irrelevant; to question
motivations, blame party affiliations, or to resort to other *ad
hominem* protections against literally *ad hominem* attack.[88] All such
responses were common enough in the seventeenth and eighteenth
centuries. They were, by the time Arbuthnot was writing, very much
a function of party commitment. Accusing a satirist of writing for
a party was used to balance and perhaps even nullify the direct and
personal attacks of the self-appointed moral censors of society. All
such defences were plausibly available to the late Lord Wharton's
friends against Swift's accusation that he had been a pathological
liar. As a corollary, ritual attacks on the party enemy preached to
the converted, and rather than reforming, might only entrench
prejudice and division.[89] This point is intimated by the Author's
warning in the 'Pseudologia' that party leaders should not believe
their own lies, which he explains in terms of 'too great a Zeal and
Intenseness in the Practice of [lying] and a vehement Heat in mutual
Conversation'.[90]

In the light of the distinction between the general and particular,
and between satire being direct or indirect in force, we can locate
The Art of Political Lying as an example of the general and in both
senses, indirect; a satire of rhetoric's open palm. It is thus doubly
appropriate that the Author asks for contributions, that we are
encouraged to flesh out the skeleton. At the very least this invitation makes a virtue of the necessity of general satire, a virtue of the
necessity of book production. As Burton had remarked, the reader
is like a guest before the author's dishes, he will sample as he will.[91]
It is appropriate to the satire's style that later readers have, as it
were, taken up the proffered dish and maintained its relevance by
intimating that the reader adjust the spices.[92] If such a style of satire
reflected an Arbuthnotian theory of satire's role and the conditions
of its likely success, it is a view Pope neither adequately confronts

nor rebuts. With only a marginal adjustment of focus the Projector's general comments on 'the Tendency of the Soul towards the
Malicious'[93] point to the dangers Arbuthnot was perhaps already
seeing in particular satire, stemming 'from Self-Love, or a Pleasure
to find Mankind more wicked, base or unfortunate, than our
selves.' Wherever it could so be seen its work of reformation was
undone.

VIII

Nevertheless, as I have said, we are not dealing with absolute satiric
dichotomies but with a tensile balance of dispositions and under-
lying the differing satiric styles of Swift, Pope and Arbuthnot, was
an imperfectly shared understanding of the manipulative function
of satiric wit and humour, one of the aspects of rhetoric which
gradually became most central to satire. If 'the satyrist loues
Truthe, none more than he',[94] he hates all vice, the archetypal
reactions to which are laughter or tears:

'Kinde pity checks my spleene; brave scorn forbids
Those tears to issue which swell my eye-lids;
I must not laugh, nor weepe sinnes? and be wise
Can railing then cure these worne maladies?[95]

Although one might dissolve into the other, it is with laughter that
satire is usually associated. Thomas More's belief that the whole
human condition was ridiculous was the reaction of a satirist.[96] But
humour was, to repeat, not a necessary characteristic of satire but
rather a means which has later been confounded with the end of
reform. Erasmus in *In Praise of Folly* and in reflecting to Martin
Dorp, on its reception made it clear that evoking laughter was
a rhetorical strategy, lengthily discussed by the ancients.[97] Burton
for example presupposes a clear difference between satire and
laughter by stating that he writes in the comic mode and then the
satirical; and Dryden distinguished them by remarking that 'satire
lashes vice into reformation, and humour represents folly so as to
make it ridiculous'.[98] Within the context of rhetoric the attempt to
provoke laughter had itself been a much theorized persuasive
device;[99] and this function is enough to explain its role in the more
aggressive forms of satire. As Samuel Butler noted, no man laughs

fully without baring his teeth; given such a bestial association, it could be held on ancient authority that the truly wise do not langh.[100] Yet, laughter could also be understood in more benign, or elevated terms.[101] Isaac Barrow and John Straight delivered sermons distinguishing Christian laughter from secular and more dubiously motivated provocations to mirth.[102] Part of Erasmus' justification for using humour as a rhetorical strategy had been to make his satire more effective by 'disarming' those he amused.[103] Whereas Addison in an early number of *The Spectator* strongly endorsed Hobbes's view that laughter is pride in one's superiority over another, he later argued in more Erasmian vein that this was only mostly true.[104] Although laughter was not possible until the pride that came with the Fall, we also regard laughter as an expression of joyfulness and amiability beautiful in itself.[105] A further claim which cemented the differing functions of laughter into the very soul of satire was advanced by the Third Earl of Shaftsbury, arguing that the provocation of laughter, largely by ridicule, was a vital means of establishing moral truth. Humour, at least if it involves true and fitting wit, befriends truth by its hostility to dogmatism.[106] Irrespective of whether laughter is associated with contempt, its enjoyment is potentially a binding and civilizing force.

Ambivalent attitudes to laughter would out-live Arbuthnot. Adam Smith's *Lectures on Rhetoric* illustrate the uncertain range of classical and Hobbesian rhetorical theories of laughter. Much hinges on a dual sense of ridicule as to laugh at, thus showing contempt for, and to be deemed ridiculous, which may be less pointed, less malign and less tied to the eristics of discourse. Thus on the one hand, in what might be a direct allusion to Hobbes, Smith disputes what some philosophers have maintained that laughter arises from and is designed to instil contempt.[107] For situations of simple incongruity (Gulliver in Lilliput) may seem ridiculous enough to make us laugh without holding any one person in contempt. Yet he goes on to equate humour with ridicule and this is extended well beyond the bounds of persuasive discourse. Any situation at which we may laugh diminishes the target of our amusement.[108] Laughing with the rhetor was laughing at an enemy; thus the successful attempt to elicit a sense of amusement was to move an audience in an appropriate way.[109] As Hobbes had recognized when digesting ancient rhetorical theories, when we laugh we achieve an instant glory, are joined with one against another.[110] Laughter could become a proof of rhetorical victory and as such it is easy to see why it has been of central importance to the satirist. It is striking that Smith's analysis of laughter is developed mainly by comparison between the satiric styles of Lucian and Swift who together, he claims, may be taken as offering a complete system of ridicule which is a system of morality.[111] To be provoked to laughter by the satirist is to be made to respond to sin.[112]

The general consequence of the eristics of laughter had been the elaboration of a complex tropology of laughter provoking devices, the satirist's fundamental tools of rhetorical trade.[113] These such as irony, *litotes*, *tapinosis* and above all, in Arbuthnot's case, *aposiopesis* are found throughout *The Art of Political Lying* as they are found in the harsher satires of Pope and Swift and before them Hobbes and Marvell. Yet in *The Art of Political Lying*, there is a lack of exclusive focus and particular victimisation as one would expect from Arbuthnot's general open-handed satire. And a more particular consequence of the recognized role of laughter in rhetoric together with more benign understandings of it may well have been to blur, as it were, the satiric styles of the venomous Pope with the more gentle Arbuthnot, the ambivalence of laughter being the mechanism allowing each to see the joke, out of the differing springs of humorous reaction.

Certainly an Horatian ethos of gentle irony pervades the whole 'solemn burlesque'.[114] The invitation, to appreciate the ridiculous rather than ridicule the contemptible and so risk being seen to be exhibiting and enjoining a distasteful pride, is consistent with the master trope of laughter being a form of *aposiopesis*, a structured lacuna offered for the reader to finish. As the authors of *Peri Bathous*, would explicate, *aposiopesis* is important enough to play a valuable part in creating genuinely awful poetry.[115] The crudest forms of this are scatological where rhyme or closure is effected by unstated taboo terms. 'Then *Nic.* loll'd out his Tongue, and turn'd up his Bumm to him [John Bull]; which was as much as to say, Kiss-'.[116] In knowing what those terms are, one is implicated in the joke of which the author remains formally innocent. Thus *aposiopesis* is a form of textual completion, its laughter the most intense and immediate sign that an audience has been moved by the argument, for that laughter is itself involvement rather than an ex post facto appreciation. It also most obviously flatters our abilities to keep up with the author. This of course, can be used effectively to isolate a specific victim; but in *The Art of Political Lying*, it plays its part in diversifying the targets of party division, making the satire far more general and widely applicable. How applicable depends upon the reader; if we are involved in the humour, we are also not immune from its range. To be told of 'your' typically English lie in a way that assumes the audience does not need details, invites a range of exemplification which hardly excludes any in an English reading audience. If it becomes exclusive it is a function of our vision, a failure fully to exploit what we are offered. To be invited, at the outset to help with the earnest Author's theory, is to establish exactly the importance of the trope. By laughing at others we might better know each other and learn to read ourselves. To allude to Shaftsbury's terms, in sharing a joke we attest to a common sense of civilisation, rather than isolating a victim or a party to be ridiculed and cut from the common herd.

As I have suggested in a different context, (ch. 2, I.), Arbuthnot's world was part of the long seventeenth century and it may help provide a final focus on Arbuthnot's satiric practice to counterpoint it with that of Hobbes. *Leviathan* is a work which in a number of respects stands behind *The Art of Political Lying*. Each text was written by a man with an elevated faith in mathematical learning, each offers a systematic vision of the political and albeit in very different ways, each has a preoccupation with philosophy or science. With respect to satire, both writers seem to share similar premises about the limitations of direct rational argument in a world of fragmented and hostile interests. Hobbes urged generally the importance of self-knowledge, *scito tiepsum*, which in a specifically satiric context can surely be taken as expressing a precondition for effective satire. The maxim affirms the importance of reading one's self, as it were, in the Swiftian glass. Arbuthnot held to a similar imperative and would entitle his last major work in the Greek, *Gnothi Seauton*, know thyself. Each, however, exhibits what might be called a different reading of the open palm of rhetoric as a means of overcoming those limitations which obscured true self knowledge. For Hobbes, effectively the open palm gathers to itself a great diversity of argumentative strategies in the Vallaesque hope that where one attack might fail another might overcome prejudice to the reformer's zeal. For Hobbes, attack is an appropriate term, rhetoric is a form of combat, symbolized on the frontispiece to *Leviathan* as comprising weapons analogous to those of physical war. Consistent with this is Hobbes's notoriously arrogant, imposing and brow-beating authorial persona. Like Milton's Satan, Hobbes emerges within his text as a brilliantly fascinating, if repellent image of frustrated power questing after authority. By contrast, Arbuthnot's reading of the open palm is the one that beckons and entraps with its feigned innocence and intimacy. The contrasting reliance on the trope of *aposiopesis* is symptomatic, for above all others it feigns relaxation of authorial control over the text. Whereas Hobbes hardly resorts to it, and then only to isolate a class of satiric victims, the catholic priesthood,[117] it is the master trope of *The Art of Political Lying*, made more effective in the alterations to the second edition.

IX

The Art of Political Lying then, is, I hypothesize, an Erasmian attempt to overcome the difficulty of seeing one's own face in satire's glass, made so formidable in a society which had cracked it into the catoptric facets of factional division, the implacable mendicant orders of Swift's *Sentiments*, the self-deluding and ambitious factions who, according to Hobbes, had caused those Civil Wars that still haunted the early eighteenth century. In a sense this encompassing and potentially reflexive aspect of Arbuthnot's satire was made more effective by the omission of party jibes in the 1727 edition (see above ch. 2, V) and which was brought out by Swift and Pope partly to establish a corpus of authentic works. The 'Preface' is much taken up with the problem of pirate editions and attributing factitious works to writers. The greater fame a writer has, they remark, 'the more such Trash he may bear to have tack'd to him'.[118] Such is the emphasis on a proper respect for what authors actually do write that it is unlikely that the alterations to the *Art of Political Lying* would have been made without Arbuthnot's blessing, if indeed they were not made expressly at his own suggestion. As these alterations are consistent with a considered understanding of satire, the latter is a reasonable hypothesis. As the edition was printed during the period in which Pope and Arbuthnot were moving in different satiric directions, it is less likely that Pope made them independently.[119] The new *Art of Political Lying* may well be seen as the polished counter-weight to what would be Pope's dramatic rationalizations for his satiric style. Fittingly, it is this version with all its enticing and adaptable generality which has been re-printed most often for later audiences.[120] In 1727, Arbuthnot was then only several years from finishing *Gnothi Seauton*, a philosophical Pascalian poem of consolation, an essay on salvation through self-knowledge of the paradox of gain through the recognition of loss. In this, the reflexive potential of his satire is fulfilled.

> But think not to regain thy native Skye
> By Towing thoughts of vain philosophy;
> Strange is the way that Leads to paradise
> Thow must by creeping mount & sinking Rise.
> Lett Lowly thoughts thy wary Footsteps guide,
> Regain thus humbly, what thow lost by pride.[121]

NOTES

1 Jonathan Swift, *The Journal to Stella*, ed Harold Williams, (Oxford: The Clarendon Press, 1948), 2 vols., 1, letter 30,19, Sept. 1711, p. 363.

2 Swift, *Ibid.*, 23, ix, 1711, p. 365.

3 Joseph Addison and Richard Steel, *The Spectator*, 19, Feb., (1712). The timing suggests that this was a sly dig at Swift, who had proposed a society for the control of English along the lines of the French Academie; Swift, *A Proposal for Correcting, Improving and Ascertaining the English Tongue in a Letter . . .* (1712). The political significance of language had been stressed by Swift. His *Proposal* had been scathingly attacked by John Oldmixon, *Reflections on Dr Swift's Letter to the Earl of Oxford* (1712) in part as a Tory plot and Frenchified fancy. Swift and Addison were not on the best of terms.

4 *The Plain Dealer*, 12, 14, July (1712); see also Lester M. Beattie, *John Arbuthnot Mathematician and Satirist*, (Camb. Mass: Harvard University Press, 1935, and NewYork: Russell & Russell, 1967) pp. 290–2.

5 Beattie, *Ibid.*, p. 292.

6 Alexander Pope, in Addison and Steele, *The Spectator*, 457, 14, Aug., (1712). This was in parody of a long running series of cribs by Samuel Parker and George Ridpath et. al. *The History of the Works of the Learned* (1699–1712). The satiric call was to be to be taken up eventually through the formation of the Scriblerian Club.

7 Pat Rogers, *Literature and Popular Culture in Eighteenth-Century England* (Sussex: Hurvester, 1985), pp. 18ff. On my rough estimate between a quarter and a third of the *The Post Boy* and *The Daily Courant* during 1712 was taken up with advertising. But for the summer months, printing advertising was a significant proportion of this.

8 Swift, *Journal to Stella*, letter 53, 9 Oct. (1712) p. 562 and n. See *The History of the Works of the Learned*.

9   Patricia Carstens, 'Political Satire, in the Work of John Arbuthnot,' London University unpublished PhD thesis (1958), p. 295, and n. 3 quoting *The Daily Courant*, 5, Sept., (1712). The seven for six offer is also found in an advertisement for the works of the Revolutionary whig Rev. Samuel Johnson, for those 'who would encourage so valuable a work', *The Daily Courant*, 3433 13, Oct. (1712). In fact during 1712 subscription advertising was rare in *The Daily Courant*. The Commynes subscription was filled and *The Daily Courant* advertised its production and delivery on 2nd and 5th of December.

10 This also was a genuine feature of subscription printing, see the advertisement for the works of Dr Johnson, *Daily Courant* 13, Oct. (1712).

11 *The Daily Courant*, 3438, 18th Oct.; 3440, 21. October, (1712).

12 Rogers, *Literature and Popular Culture*, p. 27, 8, citing Swift, *Prose* ed. Davis, vol. 4, p. 58.

13 Howard Erskine-Hill, *The Augustan Idea in English Literature*, (London: Edward Arnold, 1983), see especially the discussion of Donne's and Pope's imitations of Horace.

14 At once society is criticized and the author's moral integrity asserted through the courage to write. The problem of assessing whether fears are feigned, exaggerated, genuine let alone justifiable is much vexed. Margaret Rose, *Parody:Metafiction*, (London: Croom Helm, 1979) p. 117 has some valuable remarks on the use of animal fables in this context; Carstens, 'Political Satire' pp. 50ff for scepticism on the likelihood of persecution in the early eighteenth century; Annabel Patterson, *Censorship and Interpretation: The Conditions of Writing and Reading in Early Modern England*, (Wisconsin: University of Wisconsin Press, 1984), for a strong emphasis on the dangers of persecution; Glenn Burgess, *Absolute Monarchy and the Stuart Constitution* (New Haven: Yale University Press, 1996), pp. 5ff for scepticism in line with Carstens.

15 Carstens, 'Political Satire', p. 46.

16 Carstens, *Ibid.*, p. 50, 212.

17 Ellen Leyburn, *Satiric Allegory Mirror of Man* (Westport: Greenwood Press, 1978), ch. 1 for the general characteristics.

18 Carstens, 'Political Satire', p. 46; Bower and Erickson, eds., *The History of John Bull*, (Oxford: The Clarendon Press, 1976), p. lxxv.

19 Carstens, 'Political Satire', p. 294 n. 2; Bruneteau, 'John Arbuthnot (1667–1735) et les idées du debut du dix huitieme siècle', Doctoral thesis, Université de Lille, (1974), pt. 2, ch. 3, p. 303.

20 Bruneteau, *Ibid.*, p. 303.

21 Carstens, 'Political Satire', p. 318.

22 Claude Bruneteau, 'John Arbuthnot et les idées', pt. 2, ch. 3, pp. 302–3.

23 *APL*, p. 20; Carstens 'Political Satire', aptly notes that 'the reader must do part of the work, and must draw his own conclusions' pp. 294, 326, which as she also notes, will have its own flattering reward, p. 306. This sits oddly her insistent whiggish reductionism of the political point of the whole tract (above, ch. 2, IV, V).

24   *APL*, p. 6.

25   *APL*, pp. 16, 17; Carstens, 'Political Satire', p. 321. This was not altered in the 1727 edition which otherwise shows signs of careful correction.

26   See also Carstens, 'Political Satire', p. 336.

27   Swift *Battle of the Books*, (1710), Preface; see also *The Tale of a Tub*, (1710), Preface.

28   Peter Elkin, *The Augustan Defence of Satire* (Oxford: Clarendon Press, 1973), notes this as a prime indication of Swift's doubts about the efficacy of satire, but he goes on to claim that Swift was the only Augustan satirist who had such doubts, p. 88.

29   I am grateful to Tony Cousins for drawing my attention to *Hudibras* in this context. See also Samuel Butler, *Characters.' A Leader of a Faction'* (1667–9?) ed. C. W. Daves (Cleveland: Case Western Reserve University Press, 1970), pp. 190–1.

30   Thomas Hobbes, *Leviathan* (1651), ed. Richard Tuck, (Cambridge: University Press, 1991), Introduction, p. 10.

31   Swift, *Battle of the Books*, Preface. Modern discussions make the same point, see for example Gilbert Highet, *The Anatomy of Satire*, (Princeton: University Press, 1964), at length.

32   Samuel Johnson, *The Lives of the Poets*, 3, 61 cited in Louis Bredvold, 'The Gloom of the Tory Satirists', *Pope and his Contemporaries*, ed. J. Clifford and Louis Landa, (New York: Oxford University Press, 1949), p. 10. The strictures stopped short at Arbuthnot.

33   Sir Philip Sidney, *Apology for Poetry* (1595); John Dryden, *A discourse Concerning the Original and Progress of Satire* (1693) in *On Dramatic Poesy and Other Critical Essays* ed. S. Watson, (London: Dent, 1962), vol. 2, 71ff.

34   Peter Elkin, *The Augustan Defence of Satire*, ch. 4.

35   Anthony Caputi, *John Marston, Satirist*, (New York: Cornell University Press, 1961), pp. 24ff.

36   James Sutherland, *English Satire*, (Cambridge: University Press, 1962), ch. 1.

37   John Donne, *The Satyres, Epigrams and Verse Letters* ed. W. Milgate (Oxford: The Clarendon Press, 1967), Introduction, p. xxi; Dryden, *The Original and Progress of Satire*, on the value of ancient models, p. 137; on the importance of unity of theme for success in reform, pp. 145–6.

38   Caputi, *John Marston*, pp. 24ff.

39   David Johnston, *The Rhetoric of Leviathan: Thomas Hobbes and the Politics of Cultural Transformation*, (Princeton: University Press, 1986), has rediscovered the satiric dimension of *Leviathan*; Quentin Skinner *Rhetoric and Reason in the Philosophy of Thomas Hobbes*, (Cambridge: University Press, 1996), has taken the matter much further.

40   Elkin, *The Augustan Defence*, ch. 8, esp. p. 146.

41   Highet, whose understanding of satire *per se* is prejudiced by taking for granted the typicality of 'great' satirists such as Swift and Horace nowhere mentions Arbuthnot in his attempt to define by accumulation,

and Arbuthnot hardly fits his strictures, *Anatomy of Satire*, see for example p. 156.

42   Beattie, *John Arbuthnot*, p. 383; Erskine-Hill, *The Augustan Idea*, p. 305ff; Bruneteau, 'John Arbuthnot et les idées', Appendice C. The notion of personal satire was in practice difficult to keep distinct from satire directed at specific, ephemeral topics, but the central distinction in terms of the issue of reform is that between personal and general.

43   For an informative but unrigorous discussion see Gilbert Highet, *The Anatomy of Satire*, ch. 1.

44   Erasmus, *Morae encomium*, (1511?) *In Praise of Folly*, trans. Clarence M. Miller, (New Haven: Yale University Press, 1979), prefatory Letter to More p. 4.; see also *Letter to Martin Dorp* (1514) pp. 146–7 and usually printed with the main text; Erasmus is the seminal authority behind the notion of general satire; but see also Thomas More *Utopia* (1516) a work which praises a named individual, Cardinal Morton, but satirises only types, such as lawyers and priests.

45   Joseph Hall, *Satires*, (1597) eds., Samuel Warton and Thomas Singer, (Chiswick, 1824), xcvi, and xcii-iv.

46   Robert Burton, *The Anatomy of Melancholy*, Intro. Holbrook Jackson, (London: Dent, 1961 edn.), Intro. vol. 1, p. 121.

47   Dryden, *The Original and Progress of Satire* refers to the objects of private satire, p. 146 and avoids a choice between general and particular by advocating that individual targets be shown to be evil rather than be simply accused of wickedness. The text makes clear he has his own *Absolem and Achitophel* (1681) in mind as a model, pp. 136–7.

48   Kathleen, Williams, *Jonathan Swift and The Age of Compromise*, (London: Constable, 1959), pp. 122ff for an attempt to construct a general theory from more specific satire.

49   James Bramston, *The Art of Politicks in Imitation of Horace's Art of Poetry* (1729), p. 31.

50   Jonathan Swift and Alexander Pope, et. al. *Miscellanies in Prose and Verse* (1727), vol. 1, pref, pp. 7–8, though the discussion is not specifically concerned with satire. Elkin, *The Augustan Defence*, ch. 7 surveys the hostility personal satire generated.

51   Erasmus, *Letter to Martin Dorp*, at length.

52   Robert Burton, *The Anatomy of Melancholy*, vol. 1, p. 122–3.

53   Swift, *The Examiner*, (1710–1711), ed. Herbert Davis, (Oxford: Blackwell, 1967) 14, 9, Nov., 1710, pp. 8–13.

54   Swift, *Ibid.*, p. 8.

55   Swift, *Ibid.*, p. 11.

56   Arbuthnot, *An Epitaph on Francis Charteris*, in *The London Magazine*, April, (1732).

57   Arbuthnot (?), *A Poem Address'd to the Quidnuncki's*, (1724). Authorship is discussed in Lester Beattie, 'The Authorship of the Quidnuncki's', *Modern Philology*, 30, (1933), pp. 317ff; and is accepted also by Carstens, 'Political Satire' pp. 327ff; Alexander Ross, 'The Correspondence of John Arbuthnot', Cambridge University unpublished PhD thesis, (1956), vol.1. pp. 62–3; but see also David Noakes, *John Gay: A Profession of Friendship*, (Oxford: University Press,

1995) p. 343 for attribution of the work to Gay. The political views of the two friends seem to have been similar.

58   Arbuthnot, *A Brief Account of Mr. John Ginglicutt's Treatise concerning the altercation or Scolding of the Ancients. By the Author*, (1731).

59   The notion of a 'gothic' politics of honour is is derived from Harringtonian republicanism. See *The Political Works of James Harrington* ed. J. G. A. Pocock, (Cambridge: University Press, 1977), 'gothic' for Harrington meaning broadly feudal.

60   Beattie, *John Arbuthnot*, p. 300 and 397; see also Bruneteau, 'John Arbuthnot et les idées', pt. 5, ch. 7, p. 787.

61   Attilio Brilli, *Retorica della satira, conil Peri Bathons, O L'arte d'inchinarsi in poesia di Martinus Scriblerus*, (Bologna: Il Mulino, 1973), p. 36ff.

62   George Aitken, *Life and Works of John Arbuthnot*, (Oxford: Clarendon Press, 1892), attributes the work mainly to Pope, p. 331; but see the discussion by Beattie, *John Arbuthnot*, p. 278ff; Charles Kerby-Miller ed. *The Memoirs of the Extraordinary Life, Works and Discourses of Martinus Scriblerus*, (Oxford: University Press, 1988), Introduction, pp. 54–6.

63   Kerby-Miller, *Ibid.*, p. 55.

64   E. L. Steeves, surveys the issues of authorship, (for discussion see below Appendix B). See *The Art of Sinking in Poetry: Martinus Scriblerus' PERI BATHOUS, a Critical Edition*, E. L. Steeves, (New York: Columbia University Press, 1952), pp. xxiv ff.

65   Kerby-Miller, 'Introduction', p. 75; Bruneteau, 'John Arbuthnot et les idées', pt.1, ch. 2, p. 59.

66   Noakes, *John Gay*, p. 233ff.

67   John Gay, Alexander Pope and John Arbuthnot, *Three Hours After Marriage* (1717) ed. John Harrington Smith, The Augustan Reprint Society, 91–92 (Los Angeles: University of California Press, 1961); *A Key to the New Comedy* (1717), printed as an appendix to the play, p. 215; for discussion see Noakes, *John Gay*, p. 233ff; see also below, Appendix B. That the play was not revived, despite its wit and dramatic quality has probably most to do with the resoundingly discomforting and Mandevillian sentiments placed finally in the mouth of its main female character Mrs Townley (and seemingly endorsed by the rest of the cast). Revealed as an adulteress and generally lewd lady about town, she asks, as men want only one thing from a woman and a woman has only one thing to give, how else is she to maintain her liberty, would they sooner support her on the parish? These are sentiments remarkably close to those Shaw explores in *Mrs Warren's Profession* – with similarly controversial results even in Edwardian London. *Three Hours* was very successfully revived, however, in 1996 by the Royal Shakespeare Company. Yet the relationships between performance and canonicity are not straightforward. See Brian Corman, 'What is the Canon of English Drama, 1660–1737?', *Eighteenth-Century Studies*, 26 (1992–3), p. 307ff. *Three Hours* is not discussed.

68   Steeves, *The Art of Sinking.*, p. xlvi–xlvii suggests Pope may have published *Peri Bathous* to flush out victims for *The Dunciad*.

69 Steeves. *Ibid.*, xxvii citing Leonard Welsted and James Moore-Smythe, *One Epistle to Mr. A. Pope*, (1730) Welsted, pp. vi-vii; see also Beattie, p. 278; Ross, 'Correspondence' vol. 1, p. 67; Bruneteau, 'John Arbuthnot et les idées', Appendice C.

70 Ross,' Correspondence', vol. 2, letter 183, 17 July, (1734), p. 744; *The Correspondence of Alexander Pope*, ed. G. Sherburn, (Oxford: Clarendon Press, 1956), vol. 3, p. 417.

71 Ross, 'Correspondence', letter 185, 2 Aug., (1734), vol. 2, p. 751; Pope, *Correspondence*, vol. 3, p. 423; see Beattie, *John Arbuthnot*, and Erskine-Hill, *The Augustan Idea*, p. 307 for lucid discussions of the differences.

72 Ross,' Correspondence', letter 184, 26 July, (1734), pp. 745–6, but according to Ross post-dating letter 185; Pope, *Correspondence*, vol. 3, p. 419; Erskine-Hill, *The Augustan Idea*, p. 307. Brean Hammond offers an even later date suggesting that Bolingbroke's *Letters . . . on History* influenced Pope. No cogent reason is given to exclude Bolingboke's adapting from Pope yet Hammond's more general point that both satire and history can be forms of political theory is surely sound. See, *Pope and Bolingbroke: A Study of Friendship and Influence* (Columbia: University of Missouri Press, 1984), pp. 160–161.

73 Ross, 'Correspondence', letter 186, 25 Aug., (1734), vol. 2, p. 753. Pope's 'Advertisement' for the poem states that out of respect for Arbuthnot, the *Epistle* is not as personal as it might have been. This point is discussed in J. P. Hardy, *Reinterpretations: Essays on Poems by Milton, Pope and Johnson*, (London: Routledge, 1971), pp. 90–1; Ian Donaldson, 'Concealing and Revealing. Pope's *Epistle to Dr Arbuthnot*', *The Yearbook of English Studies*, 18 (1988), p. 182.

74 U. C. Knopflmacher,' The Poet as Physician in Pope's *Epistle to Dr Arbuthnot*', *The Modern Language Quarterly* 37 (1970), pp. 441–3, 446–8.

75 Pat Rogers,'A Drama of Mixed Feelings in the *Epistle to Dr Arbuthnot*', in *Essays on Pope* (Cambridge: University Press, 1993), pp. 93–4.

76 Francis Bacon, *The Advancment of Learning*, (1605, 1629), ed. W. G. Kitchin 1861, (London: Dent, n.d.), Bk. 2, p. 148.

77 For a fine discussion on which I am drawing here, Damian Grace 'Utopia: A Dialectical Interpretation', *Moreana*, 100, *Essays for Marc'hadour*, vol. 26, p. 273ff..

78 Erasmus, *Letter to Martin Dorp*, p. 160.

79 Lorenzo Valla, *Dialecticarum.disputationem, libri III, Praefatio, 693, Opera omnia*, (Turin, 1962) which I have discussed in ''The Rhetorical Foundations of *Leviathan*', *History of Political Thought*, 11, 4, (1990) p. 714ff; but see also Cicero, *De oratore*, ed. and trans. H. Rackham, (Camb. Mass.: Harvard University Press, 1942), II. 40. 177 for the initial *topos* for this sort of advice.

80 Burton, *Anatomy of Melancholy*, p. 20.

81 Cicero, *De partitione oratoria*, trans. H. Rackman, (Camb. Mass.: Harvard University Press, 1942, 1960) xxiii-xxiv.

82 Bacon, *The Advancement of Learning*, Bk. 2, p. 148; Hall, *Satires*, xciii-iv; Ben Jonson, 'De ornatis dignitate' *Timber* (1641).

83 Dryden, *Original and Progress of Satire*, p. 146.

84 Quentin Skinner, *Reason and Rhetoric in the Philosophy of Hobbes*, (Cambridge: University Press, 1996), pp. 426ff; Johnston, *The Rhetoric of Leviathan* (Princeton: University Press, 1986), argues that Hobbes was adjusting his arguments to a wide audience, there may be some sense in this, he was certainly aware of audience diversity, but as Skinner argues, that does not explain the highly rhetorical style of the Latin edition of *Leviathan*. It does not explain why the English version was such an expensive, and *ipso facto* audience limiting book.

85 Skinner, *Reason and Rhetoric*, pp. 428ff; also Condren, 'The Rhetorical Foundations of *Leviathan*,' pp. 716–718.

86 Skinner, *Reason and Rhetoric*, pp. 394–424, for a splendid exposition.

87 *The State Quacks, or the Political Botchers* (1715) in *The Miscellaneous Works of the Late Dr Arbuthnot*, (Glasgow, 1751), p. 163.

88 *Ibid.*, p. 16–5 where types of satire often divisive and counter-productive are discussed, not least that which is enhanced with the rhetoric of 'Slander, Calumny and Detraction' p. 164.These worries remain directly in the idiom of Erasmus.

89 Swift, The *Sentiments of a Church of England Man*, (1711), p. 91; *The Public Spirit of the Whigs*, (1714) p. 6.

90 Arbuthnot, *APL*, p. 19.

91 Burton, *Anatomy of Melancholy*, vol. 1, p. 27.

92 A.F. Pollard, *Political Pamphlets*, (London: Routledge, 1897), p. 105. Reginald Reynolds and George Orwell, eds., *British Political Pamphleteers*, (London: Wingate, 1948), vol. 1, pp. 212–3.

93 *ALP*, p. 7; contary to Bruneteau, Arbuthnot's satire involves far more than a matter of compromise with the satiric style of his friends, see 'John Arbuthnot et les idées', vol. 2, pt. 5, ch. 7, p. 787.

94 A.F.Drant, *Medcinable morall*, (1566) cited in Erskine- Hill, *The Augustan Idea*, p. 84.

95 Donne, *Satyres*, 3, p. 10; Erskine-Hill, *The Augustan Idea*, p. 86; Burton, *Anatomy of Melancholy*, p. 59.

96 Richard Pace, *De fructu*, 104, 5. I am indebted to Cathy Curtis for this point. See more generally C. M. Curtis, 'Pace on Pedagogy, Counsel and Satire,' Cambridge University, unpublished PhD thesis, (1997).

97 Erasmus, *In Praise of Folly*, p. 82 citing Quintilian, *Institutio oratoria*; *Letter to Martin Dorp*, p. 148; see also Cicero, *De oratore*, II, 54ff.

98 Burton, *Anatomy of Melancholy*, p. 32; John Dryden, *Heroic Poetry*, (1672) in *Essays*, pp. 110–111.

99 Ben Jonson, 'The Parts of a Comedy and Tragedy'; *Timber*, for an excellent discussion of laughter in the context of ancient and Renaissance rhetoric see Quentin Skinner, *Reason and Rhetoric*, pp. 198–211.

100 Samuel Butler, *Characters*, cited in David Farley-Hills, *The Benevolence of Laughter: Comic Poetry of the Commonwealth and Restoration*, (London: Macmillan, 1974), p. 10; Jonson, *Timber*, 'The Parts of a Comedy and Tragedy'.

101 Burton, *Anatomy of Melancholy*, vol 2, p. 119.

102   See Farley-Hills, *The Benevolence of Laughter*, pp. 11–12.
103   Erasmus, *Letter to Martin Dorp*, p. 148.
104   Addison, *The Spectator*, 47, 24, April (1711). For Hobbes's view, which as Skinner shows was also Quintilian's, see Skinner, *Reason and Rhetoric*, pp. 393–5.
105   Addison, *The Spectator*, 249, 15, Dec. (1711).
106   Anthony Ashley Cooper, Earl of Shaftsbury, *Characteristicks*, (1711), esp. 'A Letter concerning Enthusiasm' and 'Sensus communis'. For a succinct discussion see Stanley Green, *Shaftsbury's Philosophy of Religion and Ethics* (Ohio University Press: New York, 1967), p. 120ff.
107   Adam Smith, *Lectures of Rhetoric and Belles Lettres*, ed. C. J. Bryce (Indianapolis: Liberty Press, 1983), p. 45; see also at greater length, George Campbell, *The Philosophy of Rhetoric*, (1776); and James Beattie, *An Essay on Laughter* (1776).
108   Smith, *Lectures.*, pp. 44ff.
109   Skinner, *Reason and Rhetoric*, pp. 198–9.
110   Thomas Hobbes, *Leviathan*, ch. 6.
111   Smith, *Lectures on Rhetoric*, pp. 50–1.
112   Addison, *The Spectator*, 249 15 Dec. (1711), who wishes this happened more often.
113   Skinner, *Reason and Rhetoric*, pp. 199ff , who I am generally following here.
114   Aitken, *Life and Works*, p. 126, Arbuthnot to Swift, 19, March, (1729), the expression does not refer to *The Art of Political Lying*, but captures it beautifully.
115   Arbuthnot and Pope, *Peri Bathous*, in *A Supplement to Dr Swift's and Mr Pope's Works*, (Dublin, 1739), ch. 10, p. 282. Again, exclusive authorship is not the issue, (see Appendix B) but the work shows an easy familiarity with such rhetorical tropes; and an unselfconscious use of *aposiopesis* in *The Art of Political Lying* seems unlikely, it should have been familiar from school text books.
116   Arbuthnot, *John Bull Still in his Senses*, ch.10, p. 72.
117   Hobbes, *Leviathan*, ch. 47, where the Catholic priesthood is systematically compared to the kingdom of the fairies. One comparison, insinuating propensities to fornication is left incomplete. For a discussion see Skinner, *Reason and Rhetoric*, p. 419, who can find no other example of the trope in *Leviathan*.
118   Jonathan Swift, Alexander Pope et. al. *Miscellanies in Prose and Verse*, (1727), vol 1, p. 11.
119   Bower and Erickson draw attention to Arbuthnot's cavalier disregard for conventions of punctuation and spelling and suggest that either Pope or the printer William Bowyer made a series of corrections to the text of *John Bull*, see their 'Introduction' *John Bull*, pp. xxxiv-v. While not discounting this, there is clearly more involved in the alterations to *The Art of Political Lying*, and in these Arbuthnot's hand, or one sympathetic to the sort of satiric understanding implicit in it, seems likely.
120   See Aitken, *Life and Works*; Pollard, *Political Pamphlets*; Orwell and Reynolds, *British Pamphleteers*, reprint the 1712 edition, pp. 214–223.

The second version was also the basis for the French translation in *Traités divers Traduits de l'Anglois du Dr Jonathan Swift* (1733), pp. 247ff; clearly scant notice had been taken of Swift's and Pope's strictures about wrongful attribution.

121   Aitken, *Life and Works*, p. 442.

4 The 'Pseudologia' and Scientific Learning

'From both parents [Conradus Crambe] drew a natural disposition to Sport himself with *words*, which as they are said to be the counters of wise Men, and ready money of Fools, Crambe had a great store of cash of the latter sort.'

John Arbuthnot and Alexander Pope, *Memoirs of the Extraordinary Life, Works and Discoveries of Martinus Scriblerus*, ch. 7 p. 118.

I

It is clear from the description of it that the 'Pseudologia' is a most comprehensive and ambitious theory. Priced at fourteen shillings it was not an inconsiderable tome, eleven or twelve chapters in the first volume and a second of unspecified length. When Swift wrote to 'Stella' about *The Art of Political Lying* it was to its parodic treatment of learning that he pointed.[1] We need to ask what sort of theory it purports to be. Although Arbuthnot's projecting persona is largely uncritical of its claims about lying and about knowledge itself, Arbuthnot's own attitudes to and faith in the sciences are not far from the surface. As Ross correctly notes, a '...thread of scientific preoccupation runs all the way through his life'.[2]

The seventeenth century manifested a strong impulse towards unifying theories for different fields of discourse, and even for a unifying language of theory.[3] What Hobbes attempted for civil philosophy; Wilkins had for language; Newton for physics and Locke for the philosophy of mind. As Dunton urged in an unusually serious mood, unity of the sciences ought to be the aim of intellectual life, making all our discourse one continual syllogism; much had been done by way of abridgement, but all was still diffuse, prohibiting us from truly knowing things.[4] The belief specifically that mathematics was a key to, or model for diverse forms of learning, is in one sense Platonic; it was endorsed by Renaissance Platonism and by Galileo's claim that mathematics was the universal language of God through which we could truly know his creation. But the forms of mathematics in which there was so much faith could vary. For some, in particular Hobbes, the deductive and definitional certainties of geometry were crucial.[5] For others during the seventeenth century, notably Hobbes's critic Richard Cumberland, it was the sort of mathematics which could impose some order on uncertainty, the more inductive mathematics of probability that was so exciting. It was this, not Hobbesian science that held out the possibility of providing clear rules for politics and morals.[6]

Now the 'Pseudologia' is presented as a theoretically unifying achievement, which perhaps more surprisingly incorporates Arbuthnot's own faith in probability. The Projector tells us that it has surely won a place in the Encyclopaedia. Evocatively, the opening claims of the work are strikingly Hobbesian, even if more immediately they are the answering call to Swift's queries about the origins of lying.[7] We are loftily informed that the origins of all arts and sciences lie in scatted theorems and practices passed among initiates until some great genius comes along, 'who Collects these disjointed Propositions, and reduces them into a regular System'.[8] Hobbes had been insistent, though hardly singular in asserting that science and scientific progress was only possible by the transmission of concepts through a public shared language. He had also claimed more notoriously that political science, for all the diverse scratchings of antiquity, was no older than his own work *De cive*. The Author of the 'Pseudologia' is assuming the mantle of a similar achievement. As I have noted above, Martinus Scriblerus would do so on his own behalf in *Peri Bathous*. With the clear emphasis of litotes, the Projector tells us that the Author 'proposes to himself no small Stock of Fame in future Ages, in being the first who has undertaken this Design . . .'.[9]

II

At the risk of oversimplification, three sorts of issue are relevant to the theoretical location of the 'Pseudologia'. These concerned the role of experience, the appropriate language for scientific discourse and the putative utility of scientific knowledge. The first centred on the claims of inductive as opposed to deductive knowledge which were advanced by some of Bacon's theories of science and in particular by his contrast between theories derived from the mind as opposed to those drawn from nature. A belief in the superiority of the latter was to become orthodox, if not universal in The Royal Society. Sound theory, genuine science required first an adequate understanding of data.[10] Hobbesian deductive science was unfashionable;[11] and the dominance of induction would not be challenged philosophically until Hume's argument that the case for it was little more than an act of faith, for the only evidence was itself a matter of induction.

Arbuthnot's own scientific work was presented as orthodoxly inductive in its emphasis. Although much of it was in mathematics, his work on probability required the gathering of data to test and formulate hypotheses, rather than just illustrate principles. The gathering and quantification of demographic data from London registers of birth and death had been undertaken by John Graunt in the early Restoration and Sir Peter Pett had drawn attention to its potential significance in his *Happy Future State* (1688).[12] Arbuthnot's role in the development of such 'political arithmetic', was to develop the mathematical ratios and probabilities of survival by gender and then, as he presents his argument, to provide an explanation.[13] Arguing as Cumberland insisted,[14] from effects to causes, he urged that mathematics confirmed a providential world. Looked at in terms of statistical probability, the unlikelihood of the two sexes ending up more or less in balance in any given year, was considerable; that this equality of numbers was maintained repeatedly, despite minor variations of death rates, was prodigious. The only cause that could be hypothesized was something beyond the extraordinary pattern resulting from the interplay of demographic parameters. The problem of explaining data had been similarly central to Arbuthnot's attack on Dr Woodward's geological theories; but in this case the argument was that the theory far outstripped the data and had assumed causes which were unnecessary or which effects did not support.[15] Towards the end of his life he published his *Essay Concerning the Effects of Air on Human Bodies* (1733), a very firmly observationally based study of climate and its interactions with health, language and institutions. It was exactly the sort of empirically grounded theory for which Montesquieu would become famous. As is clear from the deductive and speculative underpinnings of this last scientific work, and was already apparent from the demolition of Dr Woodward, it would be wrong to see Arbuthnot as a sort of latter-day Aubrey, underlabouring in the collection of facts that they might be science.[16] Like Hobbes he held the deductive model of geometry as the acme of mathematical achievement.[17] Again, like Hobbes, Arbuthnot emphasized that the achievements of geometry had been made possible by public agreement on the meanings of words.[18] Most fields of knowledge could be informed by, or even reduced to some form of mathematical principles; but these however, were not necessarily *a priori* and deductive. Arbuthnot was not like his colleague Dr Cheyne an 'iatro-mathematical' physician who believed that even problems of diet could be solved by a sort of deductive mathematics.[19] Nor was he like Martinus Scriblerus, a sort of satiric synthesis of Cheyne and Hobbes, whose philosophy of ultimate causes could proceed happily without the help of experiments.[20] Arbuthnot's own work in medicine, on diet and in defence of the controversial process of inoculation paid much more attention to experience, albeit the sort of experience that could be statistically quantified. This was to reinforce probably most immediately the thrust of Cumberland's arguments about mathematics.[21] Where fields of knowledge cannot be shaped by some branch of mathematics, 'its a Sign of the knowledge of them is very small and confus'd'.[22]

There is, then, in Arbuthnot's own faith in a variable balance of deduction and observation, a model for the claims and attitudes of the Author of the 'Pseudologia'. What happens is that Arbuthnot reduces them to an unpalatable and absurd extreme. Just as he believed that knowledge should be informed by mathematics, he also held that it could not be reduced to a series of sure equations. He was no materialist, and to reiterate, the point of his mathematical work in demography was spiritual; mathematics was indeed the language of God and could reveal his workings if not his nature. It was implicitly a rebuke to the common image of the arch materialist of his age who had asserted that '. . . the Principles of naturall Science . . . are . . . farre from teaching us any thing of God's nature . . .'.[23] As the Scriblerians were to hold, it was materialist science with room only for hard data and the theories to master them that allowed the extremes of reductionism.[24] Unthreatening perhaps in its poetic Lucretian form, the modern preoccupation with the nature of things, where all might be reduced to pure materiality, was of deep concern. If Arbuthnot himself was occasionally heir to such reductionism, he endows his Author with exactly this inflated faith in theoretical rules that only needed applying to the data in order for everything to be understood. The Author has, the Projector enthuses, reduced politics to its universal principles, Rules of a political diet as it were, deduced from a philosophical theory of the soul, but needing more data, and he promises to incorporate others' suggestions to complete the task and alleviate any blemishes. As I shall show, the predominant imagery through which the whole system of the 'Pseudologia' is presented converts the principles of politics into a form of probabilistic mathematics. The common failure of lying, claims the Author, resides in ignorance or abuse of common probability.[25] The 'Pseudologia' is an ironic inversion of propositions in *The Usefulness of Mathematical Learning* and of the early essay on probability theory, illustrated as it was through card play and informed by the direct common sense that if we play cards we have to know odds. Thereby we improve our own chances and are armed against cheats. *The Art of Political Lying* evokes the Lucianic spectre of a society of cheats,[26] the 'Pseudologia' codifies and illustrates the rules of probability by which they can keep the game going for its own sake and so it arms us against the players (see below, ch. 7).

The metaphysical image of the soul as like a plano-cylindrical speculum even if contrived into fittingly off-putting jargon, was traditional and conceptually decorous.[27] Since antiquity light had been a metaphor for understanding; and in Arbuthnot's own day, the illusions light created had become a vehicle for conveying a sense of misunderstanding, delusion and false knowledge. Genuine advances in optics had thus enriched the vocabulary of social comment and indeed had been central in seventeenth-century science and broader debates about the nature of knowledge.[28] John Dunton's *Athenian Sport*, had explored several paradoxes of vision, that no colours are real, that each eye sees a different particle; and that solar rays reflected through the ether are so small as to be insensible.[29] Swift refers to the goddess of dishonesty who flies with a huge looking glass designed to dazzle and deceive.[30] The meretricious philosophical speculation given to the Author of the 'Pseudologia' may be placed in the context of such latter-day joking; but it also refracts what had been discussed seriously in *The Usefulness of Mathematical Learning*. The Author of the 'Pseudologia' suggests that one side of the plano-cylindrical speculum was created by God and reflects things as they are, the other, added later by the devil presents things, 'by the 'Rules of Catoptrics' as they are not.[31] This convex surface takes in more rays and upon it depends the art of political lying. In the *Usefulness of Mathematical Learning*, the prime example given of the capacity of mathematical principles to elucidate the world had been taken from Newtonian theories of light and the measurement of reflection and catoptric refraction; and catoptric victualers would appear as a satiric *topos*, a few years after *The Art of Political Lying*.[32] In a way that reinforces the illusion that the subject matter of the 'Pseudologia' is reducible to rigid scientific calibrations, Arbuthnot has the Projector describe rumour mongering in a way suggestive of the vocabulary of physics, and conveying the excitement of scientific discovery. In chapter nine we are informed that the Author deals with the 'Celerity' and 'Duration of Lyes'; the 'motion' of terrifying lies, which move at a 'prodigious rate, above ten Miles an hour', is different from the whisper which travels in a 'narrow Vortex, but very swiftly'. Several phenomena cannot be explained without supposing '*Synchronism and Combination*'.[33] In what then might be a further and even typical degree of self-parody, the satire achieves a Lucianic tone, which is made clearer if we turn to the issue of accessibility and the arcane languages of theory and to that of theoretical truth and usefulness.[34]

III

One of the marks of organized disciplines and the theories pertinent to them is the generation and refinement of complex specialised vocabularies which are perforce intensional and exclusive. Such language use also characterizes established professions which began to take on firmer institutional and linguistic shape in the eighteenth century.[35] As a physician Arbuthnot was at home with one such theoretical vocabulary, as a mathematician he was at home with another. However, the quest for grand unifying theories during the seventeenth century raised the question of the status of special vocabularies in an acute form. And this question occurred in the broader context of an intermittent preoccupation with the unstable and changing nature of the English language. Thomas Sprat, the much quoted advocate for scientific 'plain-tongue' also wanted to see an academy for the preservation of English. Was it possible to establish a universal theoretical language, and if so, was it desirable to do so through some form of plain discourse, or through something more arcane such as the calculus? In their different ways Hobbes, Wilkins, Cave Beck, and above all John Locke all addressed such issues.[36] Locke set out explicitly to 'break in upon the sanctuary of vanity and ignorance' built and cemented, as he saw it, with vague insignificant and abused words passing for science, and with 'hard and misapplied words . . . mistaken for deep learning . . .'.[37] There was, in short, as there remains, an incipient tension between plain- tongue and precision. When Dunton dreamed of all people engaged in one continual syllogism, he was participating in what has been something of a communal dream about the need to impose a standardizing rigour on plain speaking. In the twentieth century, the linguist Bloomfield seemed to think that as salt was NaCL, the use of the more precise term would improve our language, and no doubt the savour of our fish.[38] The Scriblerians would provide a cap for him to wear as well as men like Dunton by pretending to complain that newspaper advertisements for lost dogs were useless because they did not employ the proper technical Latin terms.

The development of the three principal professions, of lawyers, clerics and by the early eighteenth century, physicians acerbated the tensions between accessibility and precision, for the command over a specialized vocabulary was a means of maintaining an authoritative and regulated social identity. At the same time, it made the control of English additionally fraught with imponderables created by competing forms of precision. The aspect of such complex matters relevant here is that non-intensional, and inadequately defined vocabularies were unlikely to serve the rigorous needs of universal theories; while technical ones might not only be exclusive but raise suspicions of charlatanry and obfuscation. As Arbuthnot had argued explicitly in *The Usefulness of Mathematical Learning*, one consequence of the principles established by the rigorous language of geometry had been the erosion of superstition and all forms of falsehood endemic to ordinary inherited use. The almost explicit corollary is that fraudulent theories have the opposite effect.[39] God's language was not the language of popery and the truth had never harmed true religion.[40] In his earlier *Examination of Dr Woodward's Account of the Deluge*, Arbuthnot had already blended a critique of Dr Woodward's explanatory shortcomings with a satiric treatment of his inflated language and opacous jargon, all the more effective for its gentlemanly moderation.[41] It is exactly this linguistic issue that arises several times in *John Bull*. John has a love of exotic terms which seem to indicate an expertise and he is both bamboozled and impressed by Humphrey Hocus (from 'hocus pocus') and the 'hard words' of the law.[42] The legal incantations go to John's head and he fools himself that because he learns the 'hard words', he understands the law and can master its principles and the necessities of his suit. His happy sinking into the pit of the law is indicated simply by listing the legal labels of those he has to pay and what they have him pay for, confusing because aptly decontextualized: '. . . Prothonotories, Philizers, Chirographers, . . . underclerks . . . for   enrollings,   Exemplifications . . . Examinations, Filings of Words, Entries, Declarations . . . Recordats, *Nolle Prosequi's, Certiorari's . . . Supersedeas, Habeas Corpus . . . Verily,* says *John there are a prodigious Number of learned Words in this Law, What a pretty Science it is!*'.[43] Again, he is bludgeoned by the arcana of astrology when Don Diego Dismallo (The Earl of Nottingham, sounding not unlike the hapless Dr Woodward) 'went on with a Torrent of Eclyptics, Cycles, Epicycles, Ascendents, Trines, Quadrants, Conjunctions, Bulls, Bears, Goats, and Rams, and abundance of hard words, which being put together signify'd nothing. *John* all this while stood gaping and staring, like a Man in a Trance'.[44] 'It is an amazing thing', Arbuthnot had written in his *Sermon at Mercat Cross*, 'to consider how people are bantered out of their common sense with mere names and chimeras.'[45] Martinus Scriblerus was never to escape the clutches of such tenebrious word-mongering. Again, as both Brilli and Hammond emphasise, Scriblerian suspicion of science and new learning was expressed in a distrust of the language in which it was conveyed.[46]

Whatever else it is, the 'Pseudologia' is a tome of hard words, names and the suggestion of a chimera, partly presented to us in the Author's original Greek. From the jargon of the plano-cylindrical speculum to the terrifying lie (*tō phōberon*), and the prodigious lie (*tō terātodes*) it offers a tantalizing sample of an elaborate taxonomy of falsehood and universal rules for its mastery. Stirred with this are genuinely technical terms from classical rhetoric, *tō prēpon*, *tō eulōgon*, the decorous and probable, and from the Ciceronian tradition, *bonum utile, dulce et honestum*.[47] The result is to suggest a plausible confection of the charlatanical and technical, entirely consistent with the 'Pseudologia's own theories of probability and reinforced through the authoritative languages of antiquity. It would, in fact, be altogether too simple to see this specialized theoretical vocabulary as just hocus pocus, as the work of fools (the Projector and Author) who know not what they do.[48] For, we are not presented with a mere concatenation of hard words, like Dismallo's quadrants, bulls and bears, rough monuments marking

Locke's sanctuary of vanity and ignorance. On the contrary, the very deftness of illustration gives sense to classifications and the principles of their use. The wit depends much on theoretical verisimilitude, upon the odour of genuine learning, with Arbuthnot's own voice singing partly in concert with his creatures, for the advice on lying is entirely plausible, conformable to public expectations as Locke might also have put it.[49] The consequence is an intriguing balance between, on the one hand, a dismissive parody of special languages and the misuse of theory as in *John Bull*; and, on the other, Arbuthnot's own recognition of their importance in the progress of learning, through their potential for establishing principles and truths about the world. It is then, inadequate to see *The Art of Political Lying* as simply an alliance of pedantry and Grub Street;[50] it is rather an ambivalent synthesis of the claims of expertise and accessibility. It is a case of misapplied learning typical of what were to become Scriblerian preoccupations. In the suggestion of a formally instituted society for the control of lies, their coinage and circulation we have also a parodic hint, though precious little more than this, of what Sprat and most recently Swift had seriously advocated, a society for the regulation and supervision of English.[51] Swift's associating corruptions in language, cant words, enthusiastic jargon with corruptions in politics, the decline of eloquence with the rise of tyranny, may well have proved irresistible to his friend.[52] As a whole the satire fulfils the ironic potential of 'O Grub Street! *thou fruitful Nursery of tow'ring Genius's*'.[53] Arbuthnot's shifty and reflexive treatment of the claims of theory is thus an instance of his satiric style of generality and the open palm.

IV

But if good theory establishes truth, we are led to a third issue concerning the claims of grand theory. Is it enough that it be true, must it also be useful? Aristotelian philosophy had drawn a distinction between theoretical and practical knowledge which for centuries had been much recited.[54] In the seventeenth century, the attempts to promote theories had run together the true and efficacious, adapting Cicero's argument that the successful rhetorician must always combine appeals to *honestas* and *utilitas*. The adaptation was principally a matter of emphasizing the *veritas* always subsumed by *honestas*, so eliding the distinction between a moral and intellectual virtue. There was no doubt that in Bacon's vision of the new science truth would be useful; Hobbes, despite an intermittent scepticism, also stressed the useful consequences of right reasoning and such a faith had been central to Cumberland's critique of Hobbes and to Locke's whole project in the *Essay on Human Understanding*. Indeed, the utility of science seems overwhelmingly to have been taken for granted and so no distinction between science and technology was drawn. If an explanation had no obvious practical application, it was easy to conclude that it was just a matter of time before it did. In this way, the distinction between the antiquarian fact-gatherers and the theoretical scientists had to be an unstable one, sometimes no more than a distinction between the overlapping activities of a busy mind.[55] Arbuthnot was squarely within the traditions of this idiom of the rhetorical promotion of knowledge as socially beneficial. Medicine was nothing if not useful; and *The Usefulness of Mathematical Learning* maintains, that sums are good for us. As I have already stated, at times Arbuthnot appears to have been swept away by an optimism about the unlimited advantages of mathematical discovery.[56] At others, however, he does not lose sight of a distinction between mathematical *honestas* as *veritas* and *utilitas*. Although it seems very much to be part of the logic of his enthusiasm for the value of mathematics, he does not reduce all learning to mathematics nor insist that all forms of practical knowledge can be replaced by mathematics. He tries to stop well short of a naive rationalism, or as he would probably have understood it, the consequences of materialism. The sailor needs to know the conclusions of mathematics for navigation, only the scholar needs to know the truth from which they derive.[57] Further, true mathematical learning is served best by the proper study of mathematical rules, not by the creation of the cribs and compendiums Dunton would have issued into the world and which Martinus Scriblerus would do his best to provide.[58] To put the matter another way: phrase books are useful if we travel in a foreign country, they derive from a language's grammar, but we need to know that grammar to understand the language properly. The phrase book is a crib, a handbook of advice assuming we wish to do this or that; a grammar purports to lay bare the rules by which any such phrase book can come into being. There is then a place in the world for precision and the approximations of plainness.

Again if we apply this meta-theoretical distinction to *The Art of Political Lying*, we are presented with a slippage between principles and advice, between *honestas* or *veritas* and *utilitas*. It is an indeterminacy made possible by the ambiguity of the notion of a rule. Arbuthnot and his satiric personae put much emphasis on rules, the word 'Rules' or 'Rule' being used some nine times and always with a sense of discovery and enthusiastic endorsement. There are perhaps three senses, but crucially two which are run together.[59] First there are rules as constituent principles, as in Arbuthnot's notions of the rules of geometry or of reflection. Second, there are the arbitrary rules that define a game, which although they are also constitutive rules, additionally have a regulatory dimension as they define cheating. Third, there are strategic rules for success within an activity's boundaries. Rules of this third kind presuppose the other notions of a rule and regulate the options for playing most effectively. These are the rules that operate as injunctions and advice, the rules that arise from knowing the odds in gambling assuming we want to win. These are also what Arbuthnot would advertise as the 'Practical Rules of Diet'. The tension between the requirement for some codification of knowledge and its potentially deadening consequences was recognised long before Arbuthnot and the advent of the Scriblerians, even if it was to become such a vital theme of Scriberlian satire. Francis Bacon claimed that he had resisted the temptation to reduce his understanding of law to an exhaustive and intellectually overwhelming set of rules. Instead he preferred to present the reader with aphorisms to encourage a creative engagement with his text. No such qualms would inhibit that warped off-spring of the Baconian drive to knowledge, Martinus Scriberlus. In *Peri Bathous*, and in a manner very similar to the relentless reductionism of the Author of the 'Pseudologia', he would set out all the Rules of poetic composition necessary if one really wished to sink deep in ink.[60] This satiric crib provides a clear instance of how materialist science was seen by the Scriberlians to have its logical reduction in the elaboration of Rules that replaced all talent, inspiration or practice. It is all handsomely encased in the reifying image of the rhetoric box of chapter 13, in which are all the material elements of poetry. The bathetic writer need only take them out and stir them about according to the recipes provided.[61] In both *Peri Bathous* and *The Art of Political Lying*, the satire is partially dependent upon treating constituent rules and their terminology of technical description as a series of neat and simply applicable guidelines for success.

In *The Art of Political Lying* generally all that matters is a broad distinction between constituent and regulatory rules, roughly analogous to the rhetorical criteria of *honestas/ veritas* and *utilitas*. Carstens suggests that the Author of the 'Pseudologia' is concerned with principles, the Projector with practicalities.[62] She offers, however, no argument and the point cannot really be sustained. The Author, it is made quite clear, is intending both to establish the fundamental principles of pseudology and to show how useful a knowledge of them really is; he is setting down the rules in a multivalent sense. There is no distinction of voice in the claim that the 'Pseudologia' may serve 'for a Model of Education for an able Politician'.[63] The Projector concurs and disagrees once only as to what is really within the bounds of probability, namely a temporary Whig self-denying ordinance. Further, we are told it is the Author of the 'Pseudologia' who exhorts politicians and warns them, idioms of practicality and consequence not principle however pedantic.[64] In short, the whole *Art of Political Lying* conforms to the same Ciceronian conjunction of an appeal to *honestas / veritas* and *utilitas* that one finds in Bacon's cosmic advertisement for modern science and which Arbuthnot takes quite for granted and adapts, probably out of Cumberland in *The Usefulness of Mathematical Learning*.[65] What is true will be useful. In either case, the rules are posited as objective, universal; they can be understood and analysed and upon following them depends success. Arbuthnot blithely allows the Projector and Author to offer a comforting and accessible world of simple arithmetic order. The rule-monger's confidence is unbounded, and nothing, the details suggest, escapes his scrutiny. He has little to say on the regulation of prodigious lies but even here, in what might be a parodic allusion to Cicero's advice, dragons should be sizeable, storms more than a day's ride away.[66] There is even advice as to which months of the year are most suitable for which lies (allowing for the direction of the wind) which offers just a hint of a world of quadrants and trines.[67] It is, in sum, and to apply Brean Hammond's expression for the butt of Scriblerian satire, the political vision of *homo mechanicus*.[68]

The more Machiavellian possibility of meta-theory, that truth may be useless or damaging in a world of practice, was something not to be contemplated by the Author and Projector. Yet precisely this possibility is raised by Arbuthnot's satiric attitude to his creations. If this is how the world wags, then is politics useful to what should really matter? In this there is a premonition of the Augustinian withdrawal that marks the later *Quidnuncki's* (if the work is his) and the full humility of self-recognition in *Gnothi Seauton* Nevertheless, it is an assumption shared by Arbuthnot and his personae that in principle, only the principia will give rise to the reliable handbook. Contrary however, to the discrimination found in *Usefulness of Mathematical Learning*, in which Arbuthnot is speaking only for himself, he leaves his personae exposed with the untroubled rationalistic faith that the 'Pseudologia' can function as a crib. To learn and apply the crib is sufficient for a political education.[69] There is after all a book to sell, the packaged wisdom of the sophists for the politically inexperienced. If, by extension, one might now take *The Art of Political Lying* as a satiric comment on the snake oil sellers of 'How To' instant success books, a satire no less, on Oakeshottian rationalism, it was most emphatically a satiric answer to the call for a subscription for the works of the unlearned, that pointed allusion to the long-running cribologia, *The History of the Works of the Learned*. In this way, we may see *The Art of Political Lying* almost as a Scriblerian manifesto. As I have suggested, however, the ambiguity of the notion of a rule in Arbuthnot's world combined with an optimism about the utility of scientific principles made the slide into a naive rationalism very easy. Thus deliberately or not, *The Art of Political Lying* was comment on a naivite Arbuthnot had exhibited himself in his Preface to his translation of Huygens *De ratiociniis in ludo aleae*. In this way, the satire echoed similarly optimistic sentiments expressed by Cumberland, who praised Huygens' work as reducible to sure maxims and applicable everywhere.[70] In one way, to say all this is simply to emphasize the Lucianic dimension of Arbuthnot's satire. For both writers, intellectual pretension and the claims of philosophy are a fit subject for critical wit; for each, self is a fit subject for parody, and the authority of things Greek becomes questionable in the context of a modern empire. It is also, to sharpen a point made by others, that Arbuthnot satirized what he intimately knew, that in which he had invested himself by sustained study. This may help explain the difference in satiric tone from Swift and Pope, it certainly gave him considerable confidence and authority. He was, as Rogers puts it, no 'pedant manqué'.[71]

The attempt to unpack this meta-theoretical dimension of *The Art of Political Lying*, however, discloses other contextual threads. In so far as the 'Pseudologia' offers practical advice it leads to a tradition of casuistry and an adjacent sense of probability, 'probabilism'. In so far as it offers the principles of politics, it presents us with an understanding of political systems and a *reductio* of formal rhetoric founded on the probabilities of practical reasoning.

NOTES

1   Jonathan Swift, *Journal to Stella*, ed. Harold Williams (Oxford: Clarendon Press, 1948), letter 53, 9 Oct. (1712), p. 562; in contra distinction to most other modern discussions, this is also Clande Bruneteau's emphasis, 'John Arbuthnot (1667–1735) et les idées' au debut du dix huitieme siècle' Doctoral thesis, Université de Lille (1974) vol. 1, pt. 1, ch. 3, p. 101.

2   Alexander Ross, 'The Correspondence of John Arbuthnot' Cambridge University unpublished PhD thesis (1956) vol. 1, p. 28; Bruneteau, 'John Arbuthnot et les idées', pt. 2, ch. 1, pp. 166–7; and pp. 153–300 for the most detailed description of his continuing and varied scientific interests. Arbuthnot is depicted, not unreasonably, as almost a microcosm of the developing interests of the Royal Society.

3   Barbara Shapiro, *Probability and Certainty in Seventeenth-Century England*, (Princeton: University Press, 1983) at length.

4   John Dunton, *Athenian Sport*, (1707), paradox. 118, pp. 499–500.

5   For discussions of Hobbes's deductive view of mathematical and scientific understanding see e.g. J. W. N. Watkins, *Hobbes's System of Ideas*, (London: Hutchinson, 1965); Stephen Shapin and Simon Schaffer, *Leviathan and the Air Pump: Hobbes, Boyle and the Experimental Life*, (Princeton: University Press, 1985); and the essays by Douglas Jesseph, Hardy Grant and Jan Prins in Tom Sorrell ed., *The Cambridge Companion to Hobbes* (Cambridge: University Press, 1996) chs 4, 5, and 6, respectively.

6   Richard Cumberland, *De legibus naturae*, (1672) pt. 2, ch. 4, sect. 4; John Locke, *An Essay Concerning Human Understanding*, (1690), ed. A. C. Fraser, (New York: Dover, 1959), vol. 1, bk. 4, 12, 7–9.

7   Swift, *Examiner*, 14, 9, Nov. (1710), p. 9 'Who first reduced *Lying* into an Art, and adapted it to Politicks, is not so clear from History; although I have made some diligent Enquiries . . .'.

8   *APL*, p. 6 This was a theme to be taken up again in *An Essay of the Learned Martinus Scriblerus Concerning the Origin of Sciences*, in George Aitken, *Life and Works of John Arbuthnot*, (Oxford: Clarendon Press, 1892), pp. 360ff. The work develops further the intimated arrogance of scholarship. See Lester M. Beattie, *John Arbuthnot Mathematician and Satirist*, (Camb., Mass.: Harvard University Press, 1935 and New York: Russell & Russell, 1967), p. 228.

9   Arbuthnot, *APL*, p. 6.
10   See Julian Martin, *Francis Bacon: The State, and the Reform of Natural Philosophy*, (Cambridge: University Press, 1992).
11   But for some qualification to this see Noel Malcolm, 'Hobbes and the Royal Society', in J. A. G. Rogers and Alan Ryan, eds. *Perspectives on Thomas Hobbes*, (Oxford: Blackwell, 1988), pp. 43–66; also Mark Goldie, 'Sir Peter Pett, Sceptical Toryism and the Science of Toleration in the 1680s', in W. J. Sheils ed., *Studies in Church History* 21, (1984), pp. 259–60 for Pett's admiration for Hobbes and the Royal Society.
12   Goldie, 'Sir Peter Pett', pp. 266–7.
13   Arbuthnot, *'An Argument for Divine Providence, taken from the Constant Regularity in the Births of Both Sexes'* Philosophical Transactions, The Royal Society, (1710), vol. 27, p. 186; the impact in mathematics is discussed by E. Shoesmith, 'The Continental Controversy over Arbuthnot's Argument for Divine Providence', *Historia Mathematica*, 14, 2 (1987), pp. 133ff; see also Beattie *John Arbuthnot*, p. 340ff.
14   Cumberland, *De legibus*, xviii.
15   *An Examination of Dr Woodward's account of the Deluge* (1697); this is well discussed in Beattie, *John Arbuthnot*, pp. 190–6.
16   Michael Hunter, *John Aubrey and the World of Learning*, (New York: Science History Publications, 1975), ch. 1; Joseph Levine, *Dr Woodward's Shield, History, Science and Satire in Augustan England*, (Ithaca: Cornell University Press, 1977 and 1991), pp. 40–2. Interestingly, Levine argues that Arbuthnot's own *Tables of Ancient Coins*, (1727) was open to the same criticisms as Arbuthnot levelled at Woodward and that Arbuthnot did not know it pp. 240–4. This is half right. By Arbuthnot's own standards it is most unsatisfactory, an abridgement of ancient authorities and an assemblage of quasi-reliable data, but this seems to have been his own view of the work. 'I propose no reputation by (the compilation)'. See Aitken, *Life and Work*, pp. 115–6. The difference was that men like Dr Woodward proposed a great deal of reputation by their compilations.
17   Arbuthnot, *The Usefulness of Mathematical Learning*, (1701) in *Miscellaneous Works* Glasgow, (1751), p. 8.
18   Arbuthnot, *Ibid.*, p. 9.
19   E. C. Mossner, 'Hume's Epistle to Dr Arbuthnot, 1734', *The Huntington Library Quarterly*, 7, (1944), p. 144; it goes without saying that Arbuthnot would not have been dressed by the mathematical tailors of Laputa. He was more sympathetic to the empiricism of 'iatro-mechanical' medicine embraced by Mandeville, *A Treatise on Hypochondriak and Hysteric Passions* (1711). See E. G. Hundert, *The Englightenment's Fable: Bernard Mandeville and the Discovery of Society* (Cambridge: University Press, 1994), pp. 39–42.
20   On this aspect of Martinus Scriblerus see for example Beattie, *John Arbuthnot*, p. 226.
21   Cf. Cumberland, *De legibus*, pt. 2, ch. 4, sect. 4; Arbuthnot, *Mathematical Learning*, pp. 9–10; see also Locke, *Essay*, bk. 2, 26; bk. 4, 6,

15; Arbuthnot, *An Essay Concerning the Nature of Ailments and the Choice of them, according to the Different Constitutions of Human Bodies*, (1731), 'To which is added, Practical Rules of Diet' (1732 edn.); and for discussion of Arbuthnot's defence of inoculation in this context, see Bruneteau, 'John Arbuthnot et les idées', pt. 2, ch. 2, p. 184; Roy Porter, 'Was there a Medical Enlightenment in England?', *British Journal for Eighteenth Century Studies*, 5, 1, (1982), pp. 52–3 on the relationships between statistical study and medicine.

22 Arbuthnot, *The Laws of Chance*, (1738 edn.,), pp. vii–viii; cf Locke, *Essay*, bk. 4, 12, 8.

23 Thomas Hobbes, *Leviathan*, (1651), ed. R. Tuck, (Cambridge: University Press, 1991), ch. 31, p. 252.

24 Attilio Brilli, *Retorica della satira con il Peri Bathous, o L'arte d'inchinarsi in Poesia di Martinus Scriblerus* (Bologna: Il Mulino, 1973), pp. 73–81; Brean Hammond, 'Scriblerian Self-Fashioning', *The Year Book of English Studies*, 18 (1988), pp 113–14.

25 Arbuthnot, *APL*, p. 14; p. 17 where a sort of royal society is outlined for codification and the control of public lying.

26 Lucian, 'The Lover of Lies', *Works*, trans. A. M. Harmon (Camb. Mass.: Harvard University Press, 1964), vol. 3, 1–3, pp. 320–325.

27 Arbuthnot, *APL*, p. 7.

28 Richard Tuck, 'Optics and Sceptics', in Edmund Leites, ed., *Conscience and Casuistry in Early Modern Europe*, (Cambridge: University Press, 1988), esp. pp. 326ff, 244–252. There may also be an allusion Thomas Willis's medical notion of a dual soul, out of Gassendi, Willis, *Two Discourses Concerning the Soul of brutes*, (1683), cited in Hundert, *The Englightenment's Fable*, pp. 41–2.

29 Dunton, *Athenian Sport*, paradoxes, 2, 15, 41.

30 Arbuthnot, *APL*, p. 7; Swift, *Examiner*, 14, 9, Nov. (1710).

31 *APL*, p. 7; Cf Swift, *Ibid.*, p. 10.

32 *Mathematical Learning*, p. 13; *The Humble Petition of the Colliers*, et. al (1716), in Aitken, *Life and Works*, p. 317.

33 Arbuthnot, *APL*, p. 20.

34 Robert C. Steensma, *Dr John Arbuthnot, Criticism and Interpretation*, (New York: Twayne, 1979) pp. 111–118, places stress upon self-parody in Arbuthnot's works.

35 Geoffrey Holmes, *Augustan England: Professions, Sate and Society, 1680–1730*, (London: Allen and Unwin, 1982), Introduction, ch. 6.

36 For a fine general discussion, Shapiro, *Probability and Certainty* at length; Joel Reed, 'Restoration and Repression: The Language Projects of the Royal Society,' *Studies in Eighteenth-Century Culture*, 19, (1989), p. 399, esp. for the association between pure, simple language theology and nature.

37 John Locke, *Essay*, Epistle to the Reader, p. 14.

38 L. Bloomfield, *Language*, (New York: Holt, Rinehart, Winston, 1933) p. 139.

39 Arbuthnot, *Mathematical Learning*, pp. 9–10, which in this instance seems an abridgement of the view of Hobbes, possibly mediated by Cumberland and of Locke.

40   This confidence on the alliance of truth and religion is Miltonic. On
     the associations of new science and statistical mathematics with
     religious divisions, see Mark Goldie, 'Peter Pett', p. 266ff.
41   Bruneteau, 'John Arbuthnot et les idées', pt. 2, ch. 3, pp. 301-2.
42   Arbuthnot, *John Bull*, 1, ch. 7, p. 11.
43   Arbuthnot, *Ibid.*, 1, ch. 11, p. 17-18. I have much abbreviated the list.
     Bruneteau, 'John Arbuthnot et les idées', pt. 2, ch. 3, p. 302 for
     further comment on the satire on jargon.
44   Arbuthnot, *John Bull*, p. 20; cf the character of Mr. Mopus in John
     Wilson's *The Cheats*.
45   Aitken, *Life and Works*, p. 404; Beattie, *John Arbuthnot*, p. 324 on
     Defoe making similar assertions.
46   Brilli, *Retorica della satira*, e.g. p. 37f, 70-1; Hammond, 'Scriblerian
     Self-Fashioning', p. 108.
47   Arbuthnot, *APL*, p. 8, but such terms are scattered pretty evenly and
     give the impression of sustained technicality; see also Bruneteau,
     'John Arbuthnot et les idées', pt. 2, ch. 3, p. 302.
48   Cf Patricia Carstens, 'Political Satire in the work of John Arbuthnot',
     London University unpublished PhD thesis (1958), ch. 6, followed by
     Bruneteau, John Arbuthnot et les idées' for example, pt. 4, ch. 2, pp.
     686-6.
49   Locke, *Essay*, Bk. 2, ch. 32, 4.
50   Carstens, 'Political Satire', pp. 296, 318.
51   Jonathan Swift, *Proposals for Correcting the English Tongue*, (1712),
     pp. 28-34; Irvin Ehrenpreis, *Swift, The Man His Works and the Age*,
     (London: Methuen, 1967), vol 2, pp. 542-9 has a succinct discussion.
52   Swift, *Proposals*, pp. 14, 18, 13.
53   Arbuthnot, *John Bull*, p. 94.
54   Quintilian, *Institutio Oratoria*, ed. & trans. H. E. Butler (Camb.
     Mass.: Harvard University Press 1920), II, 18, 1, with a third form of
     Creative knowledge, Dante, *De monarchia*, (1310-1317?), 1, ch. 1;
     Locke, *Essay*, bk. 4, 21, 2-3.
55   For Bacon's permeating vision, see Martin, *Francis Bacon*; Thomas
     Hobbes, *Leviathan*, for example ch. 5; Cumberland, *De legibus*; Locke,
     Essay, 'Epistle to the Reader'; Bruneteau, 'John Arbuthnot et les idées',
     pt. 2, ch. 1, p. 158ff correctly stresses the emphasis on utility.
56   Arbuthnot, *The Laws of Chance*, pref. is the most obvious case.
57   Arbuthnot, *Mathematical Learning*, p. 37.
58   See e.g. *The Daily Courant*, 3474, 9, Nov. (1712) advertising John
     Hill, *Arithmetic both in Theory and Practice made Plain and easy in all
     the Useful Rules*; Myles Davies, 'The Mathematical Post', 1, 1-2, p.
     1-8, in R. George Thomas ed. *Athenae Britannicae*, (1716-1719),
     Augustan Reprint Society, (Los Angeles: University of California
     Press, 1962), No. 97. Bruneteau also stresses the Scriblerian hostility
     to scientific cribs, 'John Arbuthnot et les idées', pt. 1, ch. 3, p. 303.
59   Locke, *Essay*, for example bk. 4, 21, 3 uses 'rule' in a similarly all
     embracing fashion.
60   Francis Bacon, *Maxims of the Law* in *Works*, ed. J. Spedding, R. L.
     Ellis, D. Heath, (London, 1857) vol. 7. I am grateful to Stephen

Gaukroger for drawing my attention to this; *Peri Bathous*, ch. 3, p. 14–15; ch. 4, p. 16; ch. 14, p. 79; and ch. 15 at length for the emphasis on Rules.

61 Beattie, *John Arbuthnot*, p. 281 remarks on the literal reduction of rhetoric to a tool box; at greater length, Brilli, *Retorica della satira*, e.g. pp. 42, 73.

62 Carstens, 'Political Satire', ch. 6.

63 Arbuthnot, *APL*, p. 6.

64 Arbuthnot, *Ibid.*, pp. 11, 17, 18.

65 Cumberland *De legibus* pt. 2, ch. 4.

66 Arbuthnot, *APL*, p. 15; Cicero, *De partitione oratoria*, ed. & trans. H. Rackham (Camb. Mass.: Harvard University Press, 1942), xxi, 73.

67 Arbuthnot, *Ibid.*, p. 18.; Cf Davenant, *The True Picture of a Modern Whig*, (1701) which also refers to lies according to season, p. 6.

68 Hammond, 'Scriberlian Self-Fashioning', p. 118.

69 Such satire would have been lost on men such as Myles Davies. 'Whatever *Secrecy* or *Mistery* the *mighty ones of the Earth* make of Policy, Greatness, Power and Government, it may all be reduc'd to few Comprehensive *Maxims* of the whole, as necessary in Practice, as they be in Theory.' He immediately gives an example so empty as to nullify his innocent reductionism: 'All Policy . . . consists in a due proportion, Order and *Harmony* of the *Parts to the whole, and of both to the Head . . .*', 'The Political Post', 1, in *Athenae Britannicae*, vol. 4, p. 1.

70 Cumberland, *De legibus*, pt. 24, 4; see also Locke, *Essay*, bk. 4, 12, 7, 8. Beattie attributes, at least to the mature Arbuthnot, a highly nuanced view of mathematical learning: principles must be precise but leave much room for judgemental variation in practice, *John Arbuthnot*, p. 361.

71 Pat Rogers, *The Augustan Vision*, (London: Weidenfeld and Nicolson, 1974), pp. 226–7. In this way Rogers distinguishes him sharply from the other Scriblerians. Joseph Levine, *Dr Woodward's Shield*, pp. 239–40 sees a division of labour and fields of expertise among the group.

5 A System of Politics

D. Diego. Since worthy Persons, who are as much concern'd for your Safety as I am, have employ'd me as their Orator, I desire to know whither you will have it by way of *Syllogism, Enthymem, Dilemma* or *Sorites.*

John now began to be diverted with their Extravagance.

Though *John Bull* has not read your *Aristotles, Platos* and *Machiavels*, he can see as far into a Milstone as another: With that *John* began to chuckle and laugh, til he was like to burst his Sides.

John Arbuthnot, *Lewis Baboon Turned Honest*, ch. 3 & ch. 5.

I

Arbuthnot of course, had read his Plato, his Aristotle and could tell his syllogism from his enthymeme, while all the time he is constructing a dilemma around us. We take it for granted that we live in political systems, but to understand our habitat requires first that we have concepts of the political and of a system. These are neither self-evident nor invariant. But precisely because we do take living in a political system as a given, we may well overlook a substantial dimension of *The Art of Political Lying*. The 'Pseudologia' purports to be a systematic theory and a disclosure of a political system. It is a system of an unusual sort.

Having outlined the theoretical presuppositions which are themselves in part the object of Arbuthnot's satire, we can now focus on the specific sort of theory satirized through parody. *The Art of Political Lying* is a pastiche of the *ars rhetorica* and its claims.[1] As the 'Pseudologia' is a reduction to basic principles, so it reduces politics to persuasive language, and this becomes a political system by virtue of the field of metaphors through which the Projector describes the work. It is these related points I wish to elaborate in turn.

II

Rhetoric as a theory of politics and as a tool in politics and law has a rich and ramified history, if it also has an ambivalent reputation.

74

In antiquity the claims for the centrality of rhetoric could be divertingly extravagant and it is with such extremity that Arbuthnot presents us in a modern guise. From what is known of writers such as Gorgias and Protagoras (and much later from the arch sceptic Sextus Empiricus) it was argued that the rhetor could create and recreate reality through words. The ambiguities in the word *logos*, meaning word, statement, noun or discourse, could have been much help here and it was even held that the rhetorician was to the world of the *polis* what the magician through his words was to the world of nature, a transformer of what we know and what is real.[2] If such theories brought considerable wealth to a man like Gorgias, who could sell his theory as *technē*, they were also to have considerable explanatory power in the hands of Thucydides. They were not, however, without horrified enemies, Plato foremost amongst them and his arguments against rhetoric, much later mediated through Ciceronian understandings of technique and verbal strategy, were to surface in a text we must note in the more specific context of lying, Augustine's *De mendacio*.[3]

It was Aristotle who, in negotiating a mean between the claims of the rhetoricians and the counter-claims of Plato, produced the most profound analysis of rhetoric. His position was that although rhetoric could be abused and did involve techniques for manipulating emotion, it was above all the logic or form of discourse, predicated on the very conditions of shared experience and uncertainty that gave a sense of the *polis*. He tried to isolate three distinct genera of rhetoric, epideictic, forensic and deliberative, each differing according to social situation and *telos*, what is to be brought about.[4] The distinctions were hardly secure, but if they pointed to varying patterns of rhetorical technique, more importantly they embedded all rhetoric within the context of social conventions and situations the acceptance of which, as it were, determined what went on. The 'Pseudologia' is clearly concerned with the genus of deliberative rhetoric. The whole art of political lying is defined, we are told in chapter two (and here we are given a quotation) as '*The Art of convincing the People of Salutary* Falsehoods, *for some good End*'.[5] It is a matter of invention, *inventio* in the rhetorical sense. In a world of vitriolic public debate where notions of guilt and innocence were entangled with issues of efficacious policy, there was an ethos of trial by print, (see above, ch. 2, I–II) and so there is in what we are told of the 'Pseudologia' a judgemental, forensic undertone. If the whole *Art of Political Lying* is satire, it may as

well be called a piece of epideictic rhetoric. It has a good range of
the characteristics normally taken to identify epideictic. It is a piece
of display, an exhibition of cleverness for the delight of an audience. It is a piece of paradoxical praise;[6] yet as praise, it has an
implicit undertone of morality.[7] Moreover, it may reasonably be
seen as a ceremonial display establishing an implausible case
through ironic praise, for the purpose teaching. Gorgias would have
been proud; he wrote first on the nature of the non-existent,
Arbuthnot is promoting a non-existent book.

Be this as it may, the whole drift of Aristotle's analysis of rhetoric
was to establish it as an explanatory category of discourse and as
such give it intellectual integrity. In this way, although, like Plato,
Aristotle held philosophy to be a superior activity, he maintained
that logic was no substitute for rhetoric in understanding what went
on in the political or legal world. Rhetoric had its own logic of
probability. This is summed up by his distinction between the
syllogism and the enthymeme, albeit each an extravagance for John
Bull.[8] Like the syllogism, the enthymeme has a minor and major
premise and a conclusion; but whereas the relationships between the
components of the syllogism are necessary; the relationships of the
enthymeme draw on shared experience and lead only to judgement
as to a probable conclusion normally in the form of an injunction,
for the point of rhetorical reasoning is to get something done. Such
a view of rhetoric's structural significance for politics seems to
underpin the 'Pseudologia' and the hints of reasoning we get in *The
Art of Political Lying* are enthymemic. The recommendations and
rules are explicitly probabilistic, drawn from a shared experience of
political discourse and its effects and designed to reform practice.
The work thus has throughout it a concomitant appeal to what are
or what purport to be the common places, *loci communes* of our
assumed knowledge of the world. For example, monstrous appari-
tions tend to scare people; but the more they are used the less
effective they are, therefore do not frighten the people with the
French King or the Pretender more than once a year. More
generally the whole rationale for the use of translatory lies is
enthymemic. Lies are most effective when most plausible; where we
can draw on truth we are most credible, therefore 'The Author
exhorts all Gentleman Practitioners to exercise themselves in the
Translatory'.[9]

This is also to suggest a shift from the preconditions for rhetoric
to its effective functioning, a slippage between kinds of rule, (see

above, ch. 4, IV). Theory as a rhetoric of rhetoric, as it were, a set of strategic rules for successful argument became more important in Rome where, as Augustine was to emphasize, rhetorical training was an adjunct to success at law to which truth was often subservient.[10] The shift in emphasis between Greece and Rome should not be exaggerated. For all their attention to strategic rules, Cicero and Quintilian accepted basic premises from Aristotle. Quintilian explicitly called rhetoric a species of logic; and Cicero neatly summed up the Aristotelian emphasis on contingency and probability by calling an argument 'a probable invention to obtain conviction'.[11] Such an overlap in Greco-Roman rhetoric and a blurring of different types of rule enables Arbuthnot to intimate a seamless web of theory in the pages of the 'Pseudologia'. Nevertheless, Cicero and Quintilian had not believed, like Gorgias, that rhetoric was sheer disinterested technique, they took a high moral tone, defending rhetoric's integrity against the Platonic riposte to the Gorgian tradition. This involved them in claims as extreme in their way as those that had been made for pure rhetorical technique. The rhetorician assumed the shape of the ideal citizen, the man of wisdom and probity devoted to the truth and the public good.[12] It was to this end that they emphasized strategy and technique. In Quintilian's famous image, rhetoric was the clothing for the body of a case to be made, a truth to be established, an action to be taken; truth might have to be dressed, but it could be dressed well or ill and in different ways. More literally, how to dress, stand, how and when to elicit laughter, how to combine *honestas* and *utilitas*, these became the issues. If Aristotle suggested the rules of rhetoric predominantly in the constitutive and regulative senses of the term rule, the Romans tried to codify the rules in the regulative and strategic senses.[13]

A variable blend of Greek and Roman notions had been mediated and adapted in England from the sixteenth century, producing a veritable tradition of school learning, of which the 'Pseudologia' appears, or rather doesn't, as a singularly subversive and eccentric part. Writers such as Wilson, Peacham, Fraunce and Renoldes had all adapted Roman rhetorical theory reducing it more or less thoroughly to clear *rules* and sure *methods*, as Angel Day advertised his own work. Rainoldes and Hobbes, in their different manners, adapted Aristotelian theory to the modern world. Generally speaking, the scope of rhetoric was narrowed in this process of transmission, excluding for example, the art of memory. Yet it remained true to its classical, principally Roman origins as an

amalgam of persuasive technique and teaching. It is this modern-
ized Roman emphasis that is most parodied in *The Art of Political
Lying*, which is thus not merely a satire of hypothetical moderns, as
opposed to a pantheon of ancients. Explicitly we are told that the
'Pseudologia' is a work designed for education as well as for
political conduct and it hints at the full panoply of rules, provisos
and advice that gushed from the fonts of Rome into the small cups
of the Renaissance world.

III

Just as the scattered Greek ties the 'Pseudologia' by association to
Aristotle and the Sophists, so the Latin clearly echoes Ciceronian
priorities, even as it adjusts them. Rhetoric is about the untrue and
useful and the strategies and structures needed to keep them forever
in conjunction. So too it has a reversed image of the Roman rhetor
as a good citizen. The ideal citizen is marked by the necessary vice
of lying in what only he is pleased to call the public good. We will
find a similar reversal in *Peri Bathous*, an inverted parodic crib of
Longinus, *On The Sublime:* just as rhetoric codified the techniques
to rouse the emotions especially through its more elevated idioms,[14]
so it may be digested and applied by anyone to leave the reader cold
and bored. In these contexts the Author's claims about the 'scatter'd Theorems and Practices' that some great genius, to wit the
Author himself, brings together is itself a statement of bare faced
mendacity. And a similar claim is found in *Peri Bathous*, the genius
being Martinus Scriblerus (see also Appendix B).[15] In each case we
are in fact getting school text book stuff and an eighteenth-century
reader would not have needed telling. Lacing his text with allusions
to and verbal echoes from the traditions of systematic theory that
came from Aristotle and Cicero and had flourished in England,
Arbuthnot could have been in no doubt as to the extremity of the
claim. It is itself an instance of one of his own categories, a
'prodigious lie' (*tō terātodes*).

Understandably, this whole theoretical tradition takes its subject
matter seriously but it did not necessarily reduce the public world
to nothing more than the persuasive use of words. *The Art of
Political Lying* comes close enough to conjuring up this Gorgian
ghost, or claiming more immediately as Peacham had that the
rhetorician is almost Godlike.[16] As I shall discuss below (ch. 7, V),

the individual rhetor, the political instigator, is a true artist in his creative capacities, just as Quintilian had maintained, even if the mechanical rule book reduces invention to painting by numbers. The political world in which the artist operates is presented as a system designed to serve the development and propagation of his work. The institutions of parties, parliament and the press each has its place: the first to invent, the second to present , the third to guarantee adequate dissemination. In so far as the people are part of the political system they too have rights of lying and these are their only rights. The whole system is delineated, only by contrasting forms of discourse which seek truth. Science is not mentioned but it is made explicit with almost complete consistency that the economic (familial) and private (social) spheres operate on trust and truth. To confuse the political and non-political is, as the Author of the 'Pseudologia' indicates, both a form of *ignoratio elenchi*, an affront to an extended notion of *honestas* and is counter-productive a failure of *utilitas*. The truth will not sell. When Arbuthnot wrote, rhetoric, largely domesticated as a teachable grammar of good style still existed in its larger and more ambitious forms under a cloud of Augustinian disapproval. The power of eloquence still bred suspicion, as Locke asserted, it was an art of deceit.[17] And it was Lockean distrust which was to form a *topos* for the Scriblerian scrutiny of rhetoric.[18] There was, then, something both ironic and supremely decorous in Arbuthnot producing a tract which conforms to one of rhetoric's classical genera, whose very subject was a theory of rhetoric openly reducing the political world to the language of deceit.

IV

This brings us to Arbuthnot's own creative use of language which transforms such bald assertions of political identity into a coherent system. One of the features of his writing is the sense of metaphorical decorum and imagination that shines through. *John Bull* is so striking because of Arbuthnot's capacity to take a metaphor and elaborate and explore it so that it can become a model imposing shape on the real object of his discourse.[19] Personifications were familiar enough but in treating the family and depressingly litigious business relationships of a merchant in such intricate and allusive detail Arbuthnot created a fresh deflationary vision of European and British politics.[20] More specifically, although the metaphorical interplay between politics and marriage was well established when Arbuthnot wrote, Mrs Bull's 'Vindication' inverts the standard force of those metaphors and provides a dismissive parody of the arguments against Dr Sacheverell delivered at his trial by Whig luminaries such as Bishop Burnet (later translated as Jan Ptschirnsooker); *tapinosis* on a grand scale and effective only because of its internal consistency.[21]

The Art of Political Lying shows the same sustained metaphorical capacity,[22] the result of which is a particular model of a political system. It may be said that all understandings of a political system require some metaphor to generate them. These have most often been physical, whether mechanical or organic, or (as there is metaphorical transference between the two) a combination of both.[23] Arbuthnot's generating metaphor however, is monetary. This is entirely fitting for the reduction of politics to language. In 1712 a monetary metaphor was more contingently appropriate. Financial issues dominated politics to an unusual degree both with respect to the end of the Wars of the Spanish Succession and the Unification of England and Scotland.[24] The very wealth of England was beginning to have an impact on the understanding of political virtue. The full consequences of the establishment of the Bank of England, the development of paper money, the acceptance of the national debt as policy and the operations of the stock exchange were all quite unclear. What is now seen as economic theory was developing fast through the writings of Barbon, Petty and Davenant, and as *The Usefulness of Mathematical Learning* shows, Arbuthnot was by no means ignorant of it. There was a clear sense of a monetary system; and indeed the control over such a system had for a long time been seen as one of the hall-marks of sovereignty. The success of the system depended upon trust, a shared understanding of value even if the tokens marking value were themselves arbitrary and the system was disclosed only in activity, circulation and exchange.[25] In theory the monetary system was co-extensive with a sovereign political entity. In practice debased currencies could operate only locally, like dialects.[26] Money had, in short, many of the features which were also held to be also the features of language. This is to suggest a form of circularity which often attends well established metaphorical habits. As J. S. Peters aptly emphasizes, by the seventeenth century, the linguistic domains of money and language had become homologous interacting systems of signification. So much so, that the development of banking and of printing and then of money in the very medium associated with words could enhance and complicate the larger systems of understanding.[27]

Indeed, money had provided a field of metaphorical expansion for discourse since antiquity. In *The Republic*, the search for truth was likened to the search for gold; more recently Bacon and then Hobbes had written of discourse in terms of coinage and counters exchange and circulation. Ben Jonson wrote of custom controlling language as 'the public stamp makes the current money' and went on to warn of being 'too frequent with the mint, every day coining'.[28] Henry Neville had used gambling and card play as a metaphor for the troubles of the Commonwealth.[29] Marvell had focused a similar blend of the monetary and speculative with subordinate associations of cheating and counterfeiting coin and the market.[30] Dryden wrote of good satire in terms of the search for gold, suggestive of the Socratic search for justice.[31] Locke, echoing Bacon's 'idols of the market', had synechdochally reduced all forms of civil discourse not concerned with philosophical truth seeking to the language of the 'market and exchange', so providing a possible *topos* for Arbuthnot's image of the market exchange of politics, decidedly not concerned with the search after truth.[32] So casual and acclimatized to language had such images become that it is easy to overlook their presence when Swift remarked of the notoriously acquisitive and parsimonious Lord Wharton that his genius lay in an inexhaustible 'Fund of *Political Lyes* which he plentifully distributes'.[33] And so it remained in the eighteenth century. Bramston would later write 'Coin words: in coining ne'er mind common sense/ Provided the Originals be French' and 'When James the First, at Great Britannias helm/ Rul'd this word clipping and word coining Realm . . .'[34] Arbuthnot himself had exhibited a good deal of this easy metaphorical interplay between words and money in an early letter to his friend Dr Charlett, dated 3.x.1695. Commenting on the dearth of political news he remarked, 'politicks are so scarce they are risen cent per cent, false news like false money are only to pass with the gove[nment] for ther is no lying but upon the side of it'. Within a few lines he turns to discussing the circulation of currency and the problem of dealing with a seriously clipped silver coinage.[35] For related reasons then, money was politically metaphorically decorous; and in *The Art of Political Lying*, a monetary system becomes the means of describing a political one in a way that reduced politics to the language and rules of mathematics. Specifically, it is reduced to a fraudulent and fragile system of exchange, enlivened much, we are told, by many recent inventions and in this way is resonant with the fears of fraud which accompanied the monetary initiatives of Arbuthnot's age.

V

In the first part of *The Art of Political Lying*, the economic, in our sense, is principally part of the subject matter as the work alludes to lies concerning bribery, spending patterns and taxation. Even so, we are informed that the Author of the 'Pseudologia' has established rules to calculate the value of a lie in pounds shillings and pence.[36] This reported arithmetical claim sets the precedent for the gradual commodification of lying – and this becomes the dominant metaphorical feature of the latter part of the satire which ends by describing lies in nothing more than market terms. Parties glut the market, retailing too much of a bad commodity at once. Then the 'Pseudologia' proposes its scheme for the recovery of credit, with three months honesty giving credit for a further six months of lying.[37] Later we hear of the circulation of what others coin.[38] Because language is described through the nomenclature of trade and finance, the Author's suggestion of a formal society to regulate lying has a double resonance, suggestive both of the Bank of England and proposals, such as Swift's, as noted above (ch. 4, III), for a society to monitor language use in part because of the deleterious political consequences perceived to stem from new coinages, contractions and corruptions.[39] In a clear allusion to the less than universal operation of the currency, we are told that base coin will be good enough for Wapping.[40] As rhetoricians had long insisted, arguments have to be adjusted according to the capacities of the audience. The whole semiotics of political discourse, hand shakes, greetings and partings, 'Hugging, Squeezing, Smiling, Bowing', dim echoes all of Ciceronian injunctions about the rhetor's posture, presence and presentation, express what are called promissory lies.[41] In what is surely an allusion to Swift's remark that few lies carry their inventor's mark, the Projector tells us that it is the Author's boast that if you bring him any lie he can tell you whose image is upon it.[42] And although the very category of a proof lie is said to be *like* a proof piece of ordinance, to test the standard, in this controlled metaphorical context, the proof lie seems almost to anticipate the notion of a proof coin.[43]

In fact, the only signs of the more familiar military imagery for rhetoric are the simile with proof ordinance and the metaphor that as ministers 'use Tools to support their Power, it is but reasonable that the People should employ the same Weapon to defend themselves, and pull them down.'[44] The relative absence of what might be seen as the dominant metaphor for rhetoric's public significance is noteworthy. The imagery of conflict and war had been pervasive in ancient Rome. As Quentin Skinner has emphasized, *ornatus* referred both to embellishment and to armament and it was from rhetoric's capacity to arm the speaker in a war of words that so many portentous claims for the rhetor's significance derived.[45] As I have noted, (ch. 3, VII) the inflationary, or perhaps operatic connotations of such imagery remained evident in writers such as Valla and Hobbes. To recall a salient image, the very title page of *Leviathan* pictures the weapons of rhetoric heaped in juxtaposition to the weapons of war. Arbuthnot's relative displacement of such metaphors has a double force. It enables him to picture a political system as essentially co-operative and negotiable. His is a commercial politics of politeness, a world in which all participants are implicated and by a process of *tapinosis*, he diminished the whole significance of politics, so implying the possibilities of amelioration without sacrificing its rhetorical character. Such an intimated deflation of politics which is both inclusive and exploitative of the rhetorical common place that rhetoric is dangerous, conspires to help us laugh at ourselves as we laugh at the system.

If this is to over-read what might plausibly be hypothesized as Arbuthnot's intentions in developing the metaphors, the consequence of the dominant metaphor is nevertheless to give an altogether a less violent vision than warfare must impose on the political. Generally, one might say, it creates an inverted version of Shaftsbury's *sensus communis*, that common sense awareness of truth shared within the society of superior individuals, and which involves among other virtues, a studied politeness and a sense of the public good.[46] Arbuthnot's conspectus of the political system exhibits an ethos of calculated rule-following around probability and interest in a game that is a law unto itself played between men who are part of a special community by virtue of their patterns of exchange and their stock of credit. This ethos of probability though in one sense abstracted from Arbuthnot's own mathematical confidence, may also have presupposed Locke's brilliant account of the highly socialised grounds for probable knowledge and the appeal to *loci communes* which he saw as entailing self-knowledge, a sense of the authority of received opinion and of conditioned expectations. Probable knowledge was the result of judgement and, surely echoing Aristotle on rhetoric, the diversity of socially situated persuasive argument.[47] For Locke as for Arbuthnot there was demonstrable knowledge beyond the compass of the probable; but within Arbuthnot's political system of falsehoods there are only the lies conforming to the rules and grounds of probability and that other class which exceeds 'common Degrees of Probability'.[48]

Craig Muldrew has argued that throughout early modern England, money and credit were closely interwoven and rather than the market breaking down community relations, it was typically an expression of them and with far more emphasis being placed on trust than we assume in a post-Smithian world.[49] If he is right, Arbuthnot's choice of money and market metaphors to develop his sense of system was singularly apposite. But just as in the real monetary system the value of currency could be uncertain and subject to change irrespective of a sovereign regulating body, so an element of contingency is maintained in Arbuthnot's model. It is explicable nonetheless, and within the laws of probability, it is measurable. A degree of contingency establishes a market autonomy for politics for which accounting and mathematics, specifically statistical probability theory can provide the guiding principles. Indeed, at one stage Arbuthnot had held, like a clutch of figures as diverse as Paolo Sarpi and Philippe de Commines that what we call chance in human affairs is ignorance of causes, as it must be in a world where God is consistently seen as a first cause. For Arbuthnot, the appeal to providence, made willingly by Commines, sparingly by Sarpi, is a shift in a causative idiom.[50] Arbuthnot's famed Royal Society paper on child mortality, which had been germinating since the 1690s was finally presented as a proof for Divine Providence. It was an affirmation of an inscrutable intelligence's presence in an otherwise explicable world (see also above ch. 4, II).[51] Fortune's causes were God's work, as Sir Thomas Browne had remarked, ignorance begat the very name of fortune, 'there is no liberty for causes to operate in a loose or straggling way'.[52] As I have said, *The Usefulness of Mathematical Learning* asserted that mathematics could inform most modes of understanding and would when they were sufficiently mature.[53] It was a significantly qualified version of what he had already proclaimed in *The Laws of Chance*. In a curious premonition of the authentic voice of M. Scriblerus,

'There are very few things which we know which are not capable of being *reduc'd* to a Mathematical Reasoning.'

And,

'All the Politicks in the World, are *nothing* else but a kind of Analysis of the Quantity and Probability in casual events, and a good Politician signifies *no more* but one who is dextrous at such Calculations; only the Principles which are made use of in the Solution of such Problems, can't be studied in a closet, but acquired by the Observation of Mankind' (Italics added).[54]

In *The Art of Political Lying*, in what may have been an act of recantation, Arbuthnot creates, however ironically, a metaphorical conspectus to confirm his youthful reductive optimism. It is to this that Bramston may well be alluding in his lines about the politician who 'probability imploys/ Nor his old lies never Lies destroys' and 'New Stories always should with Truth agree,/ Or Truth's half sister, Probability.'[55]

VI

The resultant system is also strikingly secular, which is what one would expect from a deeply religious man. The hypothetical image of the soul provides an *explanans* for the political system, but thereafter religion plays little role, far less than for Machiavelli. The counterpoint between the two writers is worth elaborating, not just because Arbuthnot knew and admired at least some of Machiavelli's work,[56] but also because of Machiavelli's central position in the casuistic context that will bring us directly to the topic of lying (see below chs. 6, 7). To the Florentine religion was fundamentally important for and in civil society, the quality of a polity being very much a function of faith in its religious ethos.[57] Certainly his overall attitude to Christianity was ambivalent, as he veered between blaming the interpretation of Christianity and Christianity itself for civic weakness; but even so, it is easy and common to exaggerate the extent to which, for Machiavelli, a political domain works by rules and values contrary to a Christo-Roman moral lexicon. For the Author of the 'Pseudologia', however, politics is the creation of the Devil, who wrought the cylindrical side of the plano-cylindrical speculum to become as Swift also put it out of Genesis 'the Father of Lyes'.[58] That is, politics is a consequence of *The Fall*. This image hints at an emphasis on the dualistic nature of fallen Man, capable of some good and much mischief which Arbuthnot was to develop most strongly in his poem *Gnothi Seauton*. It was an understanding shared notably with Swift as was, eventually, the scepticism as to the capacity of rational systems of thought to overcome it.[59] Arbuthnot's Author however, is not a man of conspicuous doubt. Like the young Arbuthnot, and certainly like the Martinus Scriblerus swelling in the satiric belly of the Swiftian circle, he has a faith in Rules that give us knowledge and improve our lot in a busy world. *The Fall* is not a problem but an origin. Once the 'philosophical' argument about the plano-cylindrical Speculum has been made, nothing more is necessary but faith in the currency of a political system. The system does not serve religion and religion has no mentioned value within it, for this is a very material world – there is little more than the acknowledgement that piety and impiety provide typical problems in the political calculus, and a litotic hint from the Projector that the Author may have problems squaring his theory with the Bible and a recent sermon (discussed below, ch. 7, VIII).[60]

Given Machiavelli's alleged discovery of the autonomy of politics from religion and morality and his disdain for matters monetary, it might seem ironic that Arbuthnot presents a more consistent image of a secular, morally autonomous political system than does the Florentine. The satire provides a clear example of that sort of theory which is not presented positively in order to be promoted or defended but in order to defend or advocate something contrary. Taken from its context such a 'negative' or refracted theory may end up re-enforcing everything its proponents disliked. Not all such theory has been generated by indirect satire, though satire not seen for what it is can be woefully counter-productive. Henry Neville claimed as much on Machiavelli's behalf.[61] It has been suggested, for example, that arguments for papal infallibility were first articulated by opponents of the papacy.[62] Arguments for absolute rights of ownership later being associated with capitalism may have been advocated initially as a *reductio* of what were taken to be the untenable arguments of spiritual Franciscans about rights of use in property.[63] Arguments for parliamentary accountability may have been put forward by men only trying to ease the pressures on their elected positions, passing a buck of responsibility to their electors.[64] Here Arbuthnot simply suggests through satire a *reductio* of altogether more qualified tendencies in Machiavelli's own writings, and in neo-Machiavellian theorists of reason of state such as Justus Lipsius; and so he puts before us the secular political system for which Machiavelli was later credited.

This reductive creation of a very distinct, if morally inverted politics suggests a further dimension of the autonomy of politics thesis. Again it is only inconsistently hinted at by Machiavelli, and in a second sense it may be considered refractive. The background to this second form of refractive theory needs outlining at greater length as Arbuthnot's satire and in particular *The Art of Political Lying* is complexly related to it.

VII

Beginning with Plato's *Republic* it was often held that there was a relationship of scale between the political and personal or private world. So the political could be seen as a macrocosm, or direct magnification of something else. Plato simply asserts through Socrates that the *polis* is a larger version of the *psychē* and in order to understand the latter we may as well turn to an analysis of the former. The simile of the short-sighted man carries the whole burden of the case.[65] Again, it is claimed that the character of the *polis* is determined by the nature of its predominant moral type.[66] What we have then, is a direct, linear relationship between part and whole, individual and political. It is a thesis of ethical reflection or magnification. Perhaps the most important manifestation of this is in the Aristotelian and Ciceronian *explanans* for the generation of political society. Individuals are predominantly co-operative and social beings. The polity completes these propensities just as its generation is explained by them.

The broad basis for an alternative to a magnificatory understanding of the relationship between individual and public is to be found in Lucretius *De rerum naturae* where associability is regarded as characteristic of humanity and more acceptably in the persistent Augustinian emphasis on Man as fallen and yet capable of society.[67] An ethically reflective relationship between part and whole is questioned in an *ad hoc* way by Machiavelli. In *The Discourses*, he presents with an air of paradox the thesis that it was the vices, the ambition and greed of the Roman patricians that played a part in creating a virtuous republic. Conversely, he also claims on occasion that it is the virtues of a reformed Christianity which might in his own day have weakened civic culture.[68] Again in *The Prince*, it is urged that private virtues such as liberality may be publically damaging. It is not that private virtue *per se* has no place in politics, but that individual virtues helping to constitute that morality have no automatic authority in civic matters.[69] We have to wait until the eighteenth century before the more extreme political autonomy position is put before us. ·

In the meantime, Hobbes directly confronted the Aristotelian-Ciceronian linear thesis of sociability and polity. In an argument which was almost lovingly paradoxical Hobbes claimed through the authority of his own science that the Commonwealth was explained by individual asocial behaviour. The Commonwealth did not complete but civilized human nature. General peace was explained, even deduced by an hypothesis about individual tendencies to war.[70] This was a strong thesis of ethical refraction and this aspect of *Leviathan* was to cast a disquieting shadow over late seventeenth-century thought. Lawson queried whether individuals as Hobbes portrayed them could create a society, there had to be some, albeit imperfect, sociability; and Cumberland, whose work Arbuthnot knew, more exhaustively re-fashioned the sociability thesis against Hobbes in a way that additionally attempted to occupy the scientific high ground.[71] Like Arbuthnot, he believed that the human social and moral world could be scientifically understood through an awareness of probability. He also attempted to re-establish a necessarily linear relationship between individual virtue and public good, vice and public ill.

Shortly before Arbuthnot was to produce his satires, however, Mandeville re-stated and extended the knub of the Hobbesian paradox in *The Grumbling Hive* later to be developed in *The Fable of the Bees*, a disturbing and potent concoction of phyrronist and Hobbesian psychology and Augustinian Fallen Man.[72] To complicate matters further, however, he did so in a way that directly threatened post-Machiavellian notions of civic virtue. Where Hobbes had sought an explanation for the generation of the Commonwealth in individual asocial expectations, Mandeville asserted that it was the continuation of vices, including the vices of the civic humanist moral lexicon, that kept the hive of politics healthy. Whereas for Hobbes the Commonwealth civilized and so created and defined a public good, for Mandeville the continuation of vice sustained it. Even the most dubious and apparently anti-social motivational well-springs of human action might help people form a social bond and so allow society to jog along. This is the extreme case of ethical refraction. More immediately it ridiculed the whole civic humanist preoccupation with corruption and reform, liberty, and the energetic public spirited endeavour which maintained it. It severed the traditional Christian relationship between personal and public good so laboriously maintained by Cumberland and became something of a catalyst for the question of what constituted corruption and political morality. It was after all supremely appropriate that a world preoccupied with the ambivalent achievements of Augustan Rome, should ask questions about the virtues of a republican political system and of a Christian ethical vocabulary that grew in Roman soil.[73] The strongly refractive attitudes of *The Grumbling Hive* were reiterated in the *Female Tatler* printed 1709–10 as a direct riposte to Richard Steele's *The Tatler* and its bland confidence that in the continual exercise of all personal virtues lies the general good.[74] As Mandeville had most succinctly maintained, even burglars play their part in creating public benefit, without them the children of honest locksmiths would starve. The flippancy might have been offputting, but not sufficiently so to deter Adam Smith from developing the doctrine of the hidden hand for the final hammer blow to civic humanist virtue and for many, the most acceptable manifestation of ethical refraction in politics.

VIII

Arbuthnot's satires were all written in the aftermath of Mandeville's initial fun poking disturbance to the nest, and different aspects of these controversies can be traced through his works.[75] The *John Bull* satires clearly present a doctrine of ethical reflection. This is not principally because the relationships between John, his family, partners and clients constitute a symbolic microcosm of British and European politics. Rather, he is able to sustain this symbolic vision only because he takes a lexicon of private morality as directly appropriate in political life. The private interest of settling a law suit amicably in order to re-establish trust is an image of the public good of international peace. Because the political system is, to allude to Plato, the individual writ large, private decency should translate into public trust.

This commitment to the Plato-Cumberland continuum did not, however, stop Arbuthnot exploiting or developing his own directly Mandevillian arguments. The purpose was to ridicule the extremes to which self-serving interests could be presented as being for the public good. In *Reasons Humbly Offered by . . . the Upholders*, (1724) he would pretend to defend the rights of·apothecaries to continue peddling their appalling potions. Among the reasons are the claims that thus are genealogies rapidly augmented and that in no other way are people to be got to church. ·

The Art of Political Lying stands chronologically between *John Bull* and *Reasons Humbly offered*, but equally between a reflective and a refractive ethics. Because the political domain is the creation of the Devil it clearly is generated directly by un-ethical propensities, the desires for the malicious and the miraculous. Yet the system is by no means anti-social. Indeed, in affirming that a shared currency of lies provides a *religio*, a binding force for society, it is an imaginative denial of the truism that community must begin with the virtues of trust. Rather, by exploring the apparent paradox that a community can be sustained by the rule-bound modification of mendacity, it raises the question as to just what we do need to trust. Politicians may have to lie, but always through controlled degrees of distortion. This creates a more ambivalent relationship between truth and falsehood than initially meets the eye (see below, ch. 7, VIII; 8, IV). In these ways Arbuthnot's system may be seen as standing in contrast to Steele and *The Tatler* more sophisticatedly if less obviously than Mandeville's 'Society of Ladies' gathered together to ridicule Bickerstaff.[76] As I have shown, the political is presented as an elaborate process of rule following. Trust in the rules of 'pseudology' allows the system to prosper. Yet this does not nullify Arbuthnot's presentation of a morally refractive relationship of Mandevillian extremity.[77] In family and social matters, explains the Projector, a man has a right to expect honesty and the fair dealing of his neighbours. In politics he has no such rights. The public good rather than being a reflection of the private realm and something of genuine common concern, in the idiom of Cumberland, is nothing more than a rule bound world of solipsistic Hobbesian self-definition. What is in the public good is what each liar determines for himself. This suggests both something of a

Hobbesian state of nature and a parodic inversion of Mandeville's Hive. Private virtue is juxtaposed to public vice and part of that vice is the absence of a genuine public good. What is presented by the Author and Projector as salient features of a political system are exactly what Arbuthnot is holding up for our more critical consideration. In fact, Mandeville and Arbuthnot may be seen each in his own way, as exploring the satiric potential of the Augustinian proposition that the ubiquity of individual sinners does not prohibit some form of society and each is able to do so by presenting an image of humanity which is almost morally desiccated.[78] Where Mandeville reduced human beings to animals pursuing interest and impulse, Arbuthnot reduces them to being mechanical Rule followers in an amoral self-interested game. The two patterns of satiric critique would come together in Mandeville's Enquiry into the *Origins of Honour* with its unrelenting emphasis on dishonesty and self-seeking as the principal mechanism explaining political behaviour.[79] In a sense, one may also say that each writer developed a system of politics from *ad hominem* accusation. It was common enough to attack people by calling them liars and to accuse them of following only malodourous self-interest. For Mandeville and Arbuthnot such accusation becomes an explanatory *topos.* Yet there are salient differences in each satiric system. Mandeville's political system when fully developed in *The Fable of the Bees* would be a self-centred world of dishonest manipulation arising from and perpetuated to control society; Arbuthnot's was a self-contained world of mendacity distinct from those in which we normally live. The weight of critical emphasis also shifts; Arbuthnot presents us with a system which is morally dubious because our normal expectations do not apply; Mandeville presents us with a system used more unequivocally to hold up to critical scrutiny the relevance of those values. Whether Mandeville actually drew on Arbuthnot, however, is as unclear as the possibility that Arbuthnot was initially inspired by *The Grumbling Hive.* Textually the two doctors circle each other with tantalizing inconclusiveness. But among the important differences between them this finally is noteworthy: from the *Grumbling Hive* onwards Mandeville is sufficiently insistent on the necessity of sin and hypocrisy to place him at the very edge of satiric writing, in so far as satire is engaged in censure and reform. His is a deadly serious explanation of the means by which societies work. This is not the case with Arbuthnot who is sufficiently conventional in his faith to regard

energies spent on censure and reform as not necessarily a total
waste of time.

The balance in *The Art of Political Lying* between censure and
reform is delicate, one is suggested by allusion, the other intimated
by inclusion. By keeping the whole fraudulent system and its rules
of discourse within the bounds of plausibility; by thus seeming to
elaborate on the principles by which a system works; by inviting the
reader to apply, improve and participate, he can but increase a
sense of critical awareness (see below, ch. 8, IV). This is a type of
reform in a political world where public opinion mattered enough
to be courted. In times of desperate and ubiquitous corruption and
negligible self-knowledge, Pope would argue for desperate
measures; satiric necessity requires cutting a few from the herd, a
most Machiavellian doctrine. But as Arbuthnot may have argued to
Pope, that runs the risk of creating only scapegoats. As we are part
of the system, and a public discourse of lies withers without
listeners and disseminators, we may be better armed by knowing
ourselves (see below, ch. 7, X). As Arbuthnot's own direct foray
into public political activity had relied upon a couple of plausible
lies (see below, ch. 7, X) the force of the message might have been
partly personal after all.

If in other respects Arbuthnot's image of a political system and
the theories appropriate to its understanding strengthens the case
for his own enjoyment of self-parody and satire, it also reinforces
the necessity to turn to the context of Machiavellian and later
casuistry on which the whole satire is an indirect comment. For it
was Machiavelli's *Prince* that stood at the beginnings of the turmoil
about whether it was permissible to lie, and this was subsumed
under the more general case of casuistry itself – was it ever
allowable to break moral rules? Arbuthnot's satiric *modus operandi*,
of producing an unthreatening but enticing *reductio*, was probably
never intended to confront the power of the casuistic case. As I've
argued, his whole approach to satire was hardly naive and if its
censorious edge was dulled, in compensation it could provoke
serious thought about the rules of morality by which we can live in
political society.

NOTES

1   As the work is a pastiche rather than a specific parody, it provides evidence for Margaret Rose's critique of Frederick Jameson's claim that pastiche lacks satiric force, see *Parody, Ancient, Modern and Post-Modern* (Cambridge: University Press, 1993), p. 232.

2   W. G. K. Guthrie, *The Sophists*, (Cambridge: University Press, 1971) pp. 176ff; 204ff; Jacqueline de Romilly, *Magic and Rhetoric in Ancient Greece*, (Camb. Mass.: Harvard University Press, 1975) ch. 1.

3   For a valuable discussion, C. Jan Swearingen, *Rhetoric and Irony: Western Literacy and Western Lies*, (Oxford: University Press, 1991), ch. 5.

4   Aristotle, *Rhetoric*, trans. H. Freese, (Camb.Mass.: Harvard University Press, 1926 and 1959).

5   Arbuthnot, *APL*, p. 8.

6   T. C. Burgess, *Epideictic Literature*, (Chicago: University of Chicago Press, 1902), pp. 90ff.

7   Aristotle, *Rhetoric*, 1, ii, 3–4.

8   Aristotle, *Ibid.*, 1, i, ii; 1, ii, 8.

9   Arbuthnot, *APL*, p. 11.

10 Jan Swearingen, *Rhetoric and Irony*, p. 176ff.

11 Quintilian, *Instituto oratoria* ed. & trans. H. E. Butler, (Camb. Mass.: Harvard University Press, 1920), II. 17 42; Cicero, *De partitione oratoria*, trans. H. Rackham, (Camb. Mass.: Harvard University Press, 1942) ii, 5 'probabile inventum ad faciendum fidem'; cf. also x, 34. 'inventum' here does not connote fabrication but argumentative imagination.

12 Quentin Skinner, *Reason and Rhetoric in the Philosophy of Hobbes*, (Cambridge: University Press, 1996), p. 74.

13 For a discussion of the shift from structure to strategy, Skinner, *Reason and Rhetoric*, pt. 1, esp. 121ff.

14 Skinner, *Ibid.*, on the Roman faith in 'The Grand Style'.

15 *Peri Bathous or the Art of Sinking in Poetry* (1727) ch. 1, pp. 5–7. There is little doubt that had *The Art of Political Lying* been written two years later, the 'Author' would have been Scriblerus himself.

16 Henry Peacham, *The Garden of Eloquence*, (1593), sig AB, iii^v, see Skinner *Reason and Rhetoric*, p. 87.

17 Locke, *An Essay Concerning Human Understanding, (1690), bk. 3, 10, 34.*

18 Attilio Brilli, *Retorica della satira con il Peri Bathons, o L'arte d'inchinarsi in poesia di Martinus Scriblerus.* (Bologna: Il Mulino 1973), p. 60.

19 Lester M. Beattie, *John Arbuthnot, Mathematician and Satirist*, (Camb. Mass.: Harvard University Press, 1935, and New York: Russell & Russell, 1967), p. 90.

20 Beattie, *Ibid.*, p. 78, does point out that writers like Defoe and Leslie also used litigation as a *topos*, but he is correct in emphasising Arbuthnot's willingness to explore it.

21 Alan W. Bower and Robert A. Erickson, Introduction, *The History of John Bull*, (Oxford: Clarendon Press, 1976), p. li–lii; Patricia Carstens, 'Arbuthnot's Use of Quotation and Parody', *The Philological*

Quarterly, (1969) pp. 201–11. Cf also the sustained sexual imagery for an account of the syllogism in *The Memoirs of Martinus Scriblerus*, ch. 7.

22 '. . . whenever you *start* a Metaphor, you must be sure to *Run it down*, and pursue it as far as it can go'. *Peri Bathous*, p. 47.

23 See, for example, Raia Prokhovnik, *Rhetoric and Philosophy in Hobbes's Leviathan* (Garland: New York, 1981), p. 197 for the easy movement between such fields of imagery.

24 As late as 1975 *The Scotsman* discussed issues of devolution and Scottish independence from Westminster in the inherited metaphors of money commenting that what passes for gold in Aberdeen is 'funny money' in Newcastle, *The Scotsman* 'Editorial' 25 March, (1975).

25 Craig Muldrew, 'Interpreting the Market', *Social History*, 18, esp. 169ff.

26 I am grateful to Craig Muldrew for a copy of his unpublished paper 'Hard Food for Midas'.

27 For some provisional discussion of this notion of 'prodigal's return' see Conal Condren, *George Lawson's Politica and the English Revolution*, (Cambridge: University Press, 1989), conclusion. See J. S. Peters, 'The bank, the press and the 'return of Nature'. On currency, credit and literary property in the 1690s', in John Brewer and Susan Staves eds., *Early Modern Conceptions of Property*, (London: Routledge, 1995), p. 367ff.

28 Ben Jonson, *Timber* (1641) 'De orationis dignitate'. See also Peters, *Ibid.*, p. 367ff for other examples.

29 Henry Neville, *A Game of Picquet*, (1660).

30 Andrew Marvell, *An Account of the Growth of Popery and Arbitrary Government*, 'Amsterdam' (1677).

31 Dryden, *Of Heroic Plays* (1672) in *Of Dramatic Poesy*, ed. S. Watson, (London: Dent, 1962), vol. 2, p. 81.

32 Locke, *Essay*, bk. 3, 11, 3.

33 Swift, *Examiner*, 14, 9 Nov. (1710), p. 11.

34 James Branston, *The Art of Politicks*, pp. 9 and 28; see Timothy Dykstal, 'Commerce, Conversation and Contradiction in Mandeville's *Fable'*, *Studies in Eighteenth-Century Culture*, 23, (1994), p. 93 for the interplay between conversation and commerce in *The Fable of the Bees*.

35 George Aitken, *Life and Works of John Arbuthnot*, (Oxford: Clarendon Press, 1892), pp. 15–16; Alexander Ross 'The Correspondence of John Arbuthnot', Cambridge University unpublished PhD thesis. (1956) vol. 1, letter 3, 30 April, (1696), pp. 125–6.

36 Arbuthnot, *APL*, p. 8.

37 Arbuthnot, *APL*, p. 16.

38 Arbuthnot, *APL*, p. 18, 20.

39 Swift, *Proposals for Correcting the English Tongue*, (1712), pt. 2.

40 Arbuthnot, *APL*, p. 20.

41 Arbuthnot, *APL*, p. 21.

42 Arbuthnot, *APL*, p. 21; cf Swift, *Examiner*, 14 9, Nov. (1710), p. 11.

43 The notion of proof coin seems only to date from c.1760 (*OED*) but may well be earlier. The use of the term proof in the context of

ascertaining the authenticity of paper was already established and may have been associated with money through the development of paper money at the end of the seventeenth century.

44 Arbuthnot, *APL*, p. 10.
45 Skinner, *Reason and Rhetoric*, p. 48ff.
46 Shaftsbury, *Characteristicks*, (1711), 1, 70. On the ethos of politeness see especially J. G. A. Pocock, *Virtue, Commerce and History*, (Cambridge: University Press, 1985), pp. 47ff and 103ff; and Lawrence Klein, 'Property and politeness in the early eighteenth-century Whig moralists', in John Brewer and Susan Staves, eds., *Early Modern Conceptions of Property*, p. 221ff.
47 Locke, *Essay*, bk. 4, 25.
48 Arbuthnot, *APL* p. 15.
49 Muldrew, 'Interpreting the Market' p. 169. See also Peters, 'The book, the press,' p. 375 who emphasizes the ambivalence of credit and the strains it placed on trust; J. G. A. Pocock, *The Machiavellian Moment*, (Princeton: University Press, (1975), ch. 13.
50 Arbuthnot, *Laws of Chance*, p. vi; Paolo Sarpi, *Pensieri*, ed Gaetano and Luisa Cozzi (Milan: Einaudi, 1976) for a persistent emphasis on secular knowledge being of natural causes, an issue succinctly discussed by William Bowsma, *Venice and the Defense of Republican Liberty*, (Berkeley: University of California Press, 1968), pp. 595–8; Philippe de Commines, *Memoires*, ed. J. Calmette (Paris, 1924–5), 4, 12, 2, 86; Thomas Hobbes, *Leviathan*, (1651) ed. R. Tuck, (Cambridge: University Press, 1991), ch. 12.
51 *An Argument for Divine Providence*, p. 186; cf. Gregory, Dk. 1.2^2 Fol. B. no. 19, pp. 9–10 where the calculations and hypothesis later to be theologically developed are purely mathematical.
52 Sir Thomas Browne, *Religio medici*, (1635?) ed. Henry Gardiner, (London, 1845), sect. 18.
53 Arbuthnot, *Mathematical Learning*, p. 9, 21–22.
54 Arbuthnot, *Laws of Chance*, pp. vii–viii.
55 James Bramston, *The Art of Politicks in imitation of Horace's Poetry*, (1729), p. 18, 36.
56 Machiavelli is praised as one of the fathers of political arithmetic *Mathematical Learning*; Arbuthnot owned Henry Neville's translation of Machiavelli's works; see, *Arbuthnotiana,* 'Introduction' Patricia Koster, Augustan Reprint Society, 154 (Los Angeles: University of California Press, 1972), catalogus item 107.
57 Machiavelli, *Discorsi*, (Milan: Feltrinelti, 1973), 1.9–14.
58 Arbuthnot, *APL*, p. 7; cf Swift, *Examiner* 14, Nov. 9, (1710), p. 8.
59 Kathleen Williams, *Jonathan Swift and the Age of Compromise*, (Lawence: University of Kansas Press, 1959), p. 72.
60 Arbuthnot, *APL*, p. 9.
61 Henry Neville, *The Works of the famous Nicholas Machiavel*, (1675), A Letter in Defence of Himself, no pagination.
62 Gordon Leff, *Heresy in the Middle Ages*, (Manchester: University Press, 1967), vol. 1, p. 121; Brian Tierney, *The Origins of Papal Infallibility*, (Leiden, 1972) p. 125ff.

63   Conal Condren, 'Rhetoric, Historiography and Political Theory:
      Some Aspects of the Poverty Controversy Reconsidered', *The Journal
      of Religious History*, 11 (1982), pp. 15–34.
64   Brian Tierney, *Religion, Law and the Growth of Constitutional
      Thought, 1150–1650* (Cambridge: University Press, 1982), ch. 2.
65   Plato, *The Republic*, trans. Paul Shorey, (Camb. Mass.: Harvard
      University Press, 1970 edn.) bk. 1–2.
66   Plato, *Ibid.*, bk. 9.
67   Lucretius, *De rerum naturae*, trans. L. Johnson, (London: Centeur,
      1963) s.926
68   Machiavelli, *Discorsi*, bk. 1, 3–4.
69   Machiavelli *The Prince*, ed. & trans. Russell Price and Quentin
      Skinner, (Cambridge: University Press, 1988) chs. 15–17.
70   Hobbes, *Leviathan*, chs. 12, 13.
71   *Arbuthnotiana*, ed. P. Koster, item 374. For Lawson see *An Examin-
      ation of Mr. Hobbs, His Leviathan* (1657); It had also been a major
      anti-contractarian theme of Filmer's much earlier *Patriarcha* that
      brutish man could not provide an explanation for the origins of
      society.
72   Bernard de Mandeville, *The Fable of the Bees: Or Private Vices Public
      Benefits*, (1714, amended until 1732) ed. F. B. Kaye, (1924, Indiana:
      Liberty Press, 1988) 2 vols. This is not to claim that Mandeville
      merely vulgarized or adopted Hobbesian theories. On some of the
      differences between Hobbes and Mandeville however, see Hundert,
      *The Enlightenment's Fable: Bernard Mandeville and the Discovery of
      Society*, (Cambridge: Uniersity Press, 1994), p. 66, 177ff.
73   Erskine-Hill, *The Augustan Idea in English Literature* (London:
      Edward Arnold, 1983); Shelley Burtt, *Virtue Transformed: Political
      Argument in England, 1688–1740*, (Cambridge: University Press,
      1992); more generally, Pocock, *The Machiavellian Moment*, ch. 14.
74   Hundert, *The Enlightenment's Fable*, p. 5.
75   It is unlikely that the men did not know each other being physicians
      and each having an interest in science and the theory of medicine. It
      is also possible that Arbuthnot knew what Mandeville had written.
      Although Mandeville was later to come out in clear Whig colours, his
      hostility to Steele's moral platitudes may also have provided a shared
      interest. Each doctor was also a modern on the issue of inoculation.
76   Hundert, *The Enlightenment's Fable*, p. 5.
77   *Memoirs of Martinus Scriblerus*, in which it is held that the unthinking
      individual members of a corporate system compose one thinking
      system. See Beattie, *John Arbuthnot*, p. 270.
78   Hundert, *The Enlightenment's Fable*, pp. 37–9.
79   *An Enquiry into the Origins of Honour, and the Usefulness of Chris-
      tianity in War*, (1732) Introduction M. M. Goldsmith, (London:
      Frank Cass, 1971); see Hundert, *The Fable's Englightenment*, p. 23.

Part II

6 Casuistry

'But tell me, old Boy, hast thou laid aside all thy *Equivocals* and *Mentals* in this case?'

John Arbuthnot, *Lewis Baboon Turned Honest*, ch. 4.

I

Casuistry has had a bad press, ridiculed on stage, feared for subverting morality, and in a Protestant world indelibly stained with Jesuitry and Machiavellianism, it was nevertheless, as it still remains, a dimension of any serious moral reflection.[1] Only a little needs to be said about it here as a preliminary for the more specific context of the casuistry of lying. In fact, because of casuistry's reputation for diminished moral responsibility, it has been too easy to simplify the principal theories of ethics into consequentialist and deontological, with Bentham and Kant respectively being the names around which such theories gather and with casuists being seen only as amoral consequentialists.[2] This notwithstanding, casuistry constitutes a third dimension to ethical theorizing in its own right. In ethics there is effectively a triadic relationship between deontological principles, consequences and cases.[3] As Kant, among many others, reacted against the casuists's emphasis on cases, so Bentham reacted against Kant's emphasis on principle.[4] There is a parallel here with general and particular satire. With respect to satire and ethics, there is an oscillation between delineating concepts; and we may expect that a man like Arbuthnot, given to Realist generality of satire, would also have shown greater faith in general principles than in individual cases and so have had little truck with casuistry. Broadly speaking this is so, but not quite broadly enough for it to constitute the whole truth and nothing but the truth.

II

In the sixteenth and seventeenth centuries, the archetypal examples of absolute moral principles were the Ten Commandments and the generally complementary New Testament injunctions to maintain faith and hope, to love, to charity and obedience. It was through, and with reference to them, that the moral discourse of censure and approbation was organized. Faith in an understanding of them could give confidence to satirist and sermonist alike. All such principles, however, were abstractions from moral conduct arising from and presupposing social activity. Although they were sufficiently general to encompass a wide variety experience they were inevitably compromised in practice, kept alive by adaptation. Pride might be a sin, but what counted as pride depended on cases. As Pope, the satiric nominalist had put it with respect to satire, 'Examples are pictures'.[5] The command 'Thou shalt not kill' did not prohibit executions or wars. Animals were eaten in clear soup with a clear conscience for the paradigmatic example of killing was provided by Cain and Abel, not Abel and a sacrificial sheep. Yet, as casuists realised, abstract rules are never sufficiently determinant; they are not a substitute for, or adequate synopsis of, morality. Like the law, moral teaching was developed between cases, contexts and principles.[6] When they mattered moral issues were rarely so straightforward. It had been such a weakness in Arbuthnot's realist satire that Pope had exploited. On their own, abstract principles are 'obscure, misty and uncertain . . .'.[7] They are but an abstraction from conduct, and lead back to it. To put the matter again in terms of types of rule, they were not constitutive of moral practice, they were rules of conduct to help guide it – thus the continuing relevance of Meno's paradox; to know the general principle, the concept or paradigm of a vice or virtue we have to understand cases, features and manifestations; but these presuppose the rule in terms of which they may be grouped together.[8] As a consequence, in practice there must be the possibility of the interplay between principle and case. As this is likely to be debated only where there are genuine moral dilemmas, and as there may be different principles relevant to the case, any moral injunction giving us specific counsel, may in turn be only probably sound.[9] Again, as in law, a notion of reasonable doubt comes into play. In Jesuit casuistry, it was enough that there be a respectable authority, giving a probably correct view on which one could rely for guidance. Even a minority view could count as probably sound.[10] In the case divinity of the Protestants such as Donne, Taylor and Baxter, the recourse to authority was more doubtful. Yet a sense of conscience, and context still produced in an acceptance that in hard cases on which advice was needed, there might only be probably correct responses if apparently conflicting rules could not be reconciled.[11] In a complex moral world 'probabilism' can be a courageous recognition of the inadequacy of easy answers, a refusal to delegate the difficulties of moral judgement. Conversely, the embrace of 'probabilism' can also potentially express a moral nihilism and a self-defeating scepticism. It may have been this problem at the junction of principle and individual case that Arbuthnot had in mind in alluding to the role that mathematics, specifically mathematical probability theory, might play in morality; the result could only be to give a secure basis for probabilism.[12] Be this as it may, the interplay between principle and case lies behind *The Art of Political Lying*, written when two forms of moral fear had come to circle each other in endless semi-confrontation. On the one hand, was the concern that granting exceptions to the rule would compromise the authority of the principle;[13] and on the other, that without the possibility of such exceptions, without some court of moral equity, the rules would become absurd, irrelevant and lead to moral obscenity.[14] Thus both casuists and their enemies feared for a genuine morality. So too had Arbuthnot and Pope feared for genuine satire.

III

The casuist, however, had two discernible strategies of argument, both of which are relevant to understanding Arbuthnot's satiric treatment of casuistry. The first was to accept that, under some circumstances, the imperative of a given rule, say 'thou shalt not kill' could be put to one side. This immediately raised the problem of finding a criterion of judgement by which we might know when this could be and who should decide the matter. For without some criterion, there was nothing to stop the rule being waived capriciously to promote naked self-interest. Here arose the importance of the distinction between ends and means which was to make a lot of casuistry strongly consequentialist, pulling it towards utilitarianism. The mindless or mechanical application of an abstract principle could well be counter-productive. Conduct must be judged by the ends it served; ends which might be a range of disproportionate consequences, or a higher principle than the one immediately being applied. Gradually through the Reformation there had emerged a clear difference between Catholic and Protestant casuists. For the former it was a church authority which decided and was promoted as an arbiter of ends and legitimate means; for the latter it was one's own conscience.[15] It was partly within this context that Hobbes was accused of casuistry – caught between the Scylla and Carybdis of Catholic and Protestant idioms. He was seen as advocating obedience only in as much as as it was in one's interest and also of setting up Leviathan as a substitute for morality, so creating a civil pope.[16] But in either Catholic or Protestant idioms of casuistry, there was a tendency to elaborate ever more complex rules governing the isolation of legitimate exceptions to moral commands;[17] and to reify the appeal to conscience as a principle almost independent of cases.[18] Paradoxically, at the meta-level, casuistry exhibited the rationalistic propensities to rule- mongering to which, at the level of moral conduct, it was opposed, generating distinction after distinction to minimize the contingency involved in an appeal to context, cases, consequences and conscience.[19] This was not lost on the enemies of casuistry, Protestant and Catholic, and made casuists highly susceptible to parody and accusations of insincerity.[20]

Casuistry was derided as an elaborate language game designed for freeing the villain from any sense of moral obligation.[21] It was thus reduced to moral sophistry. At one point in Wilson's *The Cheats*, the puritan minister Scruple is offered a lucrative living which his principles should require he reject; but he covets it and so reaches mentally for his casuistic rule-book. This enables him to claim the living, not least on the principle that as all moral injunctions are only probable, all are equally safe.[22] Little wonder Gabriel Daniel complained that rather than addressing seriously the arguments of Laymann and Suarez, the enemies of casuistry set up a 'meer scarecrow, a man of straw'.[23]

This dimension of the critique of casuistry, pretty well ascendant by the first decade of the eighteenth century, is clearly present in *The Art of Political Lying*, which treads lightly over a caricatured corpse. What we have are glimpses of a work preoccupied with rules and distinctions and probabilities and lacking any sense of morality as it is normally understood. The moral principle, 'thou shalt not bear false witness', is both reversed and constrained in a set of rules and distinctions which presupposes that everything is a lie and everything must be done in self-interest. Public interest is but the projection of a liar's judgement and is nothing more than a balance of probabilities. As there is no authority beyond the individual artist, there is no formalized guide to conscience, it is a very Protestant world of casuistry which is presupposed. Again, in out-lining the Author's arguments in favour of the people being allowed to lie, the Projector puts forward an argument from sheer necessity, exceptional casuistry and almost from the mouth of The First Mrs Bull. The people sometimes have no option but to lie to save themselves. Only a generation before Clarkson had argued seriously that lying was a part of casuistic, and by implication, Catholic morality, there was no place for truth either in heart or words.[24]

IV

The second and more effective strategy available to the casuist was to re-describe the exception to one authoritative principle in terms of adherence to another and if possible reconcile the competing rules, hence ideally maintain a constructive dialectic between case and rule. The doctrines of mental reservation and equivocation were designed to reconcile the demands of truth telling with other moral rules.[25] If I am being interrogated unjustly I might make use of them. If stealing was regarded as sinful, an act of theft might be mitigated under the right circumstances by being described as an act of parental responsibility. Thus rather than casuistry being about exceptional cases to a rule, it was about making them extensions of another rule, such as justice or charity.[26] As such, extensive casuis-tic arguments addressed and explored the entailments of the whole spectrum of human offices, those compounds of responsibilities and correlative rights that in their various patterns constituted the sense of moral identity – the entailments of the offices, for example, of parenthood, service and rule.[27] Extensive casuistry was powerful as it met unbending deontologists on their own terms. Principles were not compromised, the issue was which principle was most relevant, behaviour under which office. In so far as one could chose between principles and offices in the moral lexicon (parents were usually spouses and often children), it led to a form of 'probabilism'. This strategy also eased the necessity to elaborate meta-rules of conduct for identifying exceptions to a moral principle. It required instead, however, the generation of elaborate re-descriptive vocabularies as insulation against accusations of moral turpitude. Sir Thomas Browne captured much common sentiment when he counselled against putting 'new names or notions upon authentick virtues and vices'; and this is exactly what the Reverend Mr Scruple does as Wilson's emblem of casuistry in *The Cheats*. An irksome promise is redescribed as a restriction on natural liberty.[28] As Charles de Condren remarks, if necessary one is permitted to conceal the truth, 'under some figure of which the most common are hyperboles, Ironies, Amphibologies, Ambiguous words or terms of double meaning . . . and equivocations . . .'[29] In the *Reasons Humbly Offered*, as I have noted above (ch. 5, VIII), Arbuthnot was to have grim fun with such dexterity, allowing his Upholders to re-describe poisoning patients as a means of getting them to church.[30] But lest we should be distracted by thinking casuistry trivial because of an effective satiric trivialisation, it should be noted that the clearest case is that of self-defence. A duty to or right of self-defence, largely unquestioned at the level of individual conduct, could be used to re-classify accusations of rebellion and armed rising against authority. And, in this way a very large portion of what we call the political theory of early modern Europe was casuistic, from Grotius 's theories of war to Locke's and Sidney's on politics.[31] Arbuthnot's general hostility to this form of casuistic reasoning can be seen most clearly from The First Mrs Bull's 'Vindication' of the Right to Cuckoldom.[32]

The Sacheverell trial had been a clash between casuistry and principle. In an unqualified way Dr Sacheverell had preached that as monarchs were anointed by God, it was never right to overthrow them. The clear implication being that the 1688 Revolution was immoral, the present regime illegitimate. Part of what was at stake in the whole controversy was how to describe what had happened when James departed and William and Mary ascended the throne, and in 1710 there were still a number of competing accounts. Sacheverell was uncomfortable with descriptions in terms of defence, tyranny and necessity, providence and just war and indeed, the issue of present legitimacy was not as dominant or as explicit in his sermons as it was taken to be. Nevertheless, at his trial Sacheverell had disingenuously argued that he was not talking of that particular case but only elucidating apostolic principles;[33] a somewhat nugatory defensive move. His accusers, and in this context notably Bishop Burnet, argued that there were exceptions to the requirement of obedience. The relationship between ruler and ruled was essentially consensual and contractual; that if a ruler behaved tyrannically or in some other abusive manner, there remained a residual right of defence. Grotius for example allows it, and although after a period of civil war, an emphasis on passive obedience is understandable, it cannot, maintained Burnet, be elevated to an absolute principle.[34] It will only let in Papists.[35] Such a limitation to passive obedience had been clear in 1688, claimed Burnet, a when the right of self-defence had been exercised. Moreover, acceptance of this right was the authentic doctrine of the Church of England.[36] Behind such arguments lay the casuistic rhetoric of Locke, Sidney, Marvell, Milton and Lawson.

Mrs Bull, seeming to allude to the language of these writers, is made to express a succinct parodic *reductio* of this revolutionary casuistry as it was presented at Sacheverell's trial. Patricia Koster has evidenced how closely Arbuthnot follows and ridicules the trial speeches of Walpole, Lechmere and Stanhope creating a parody of 'Rabelaisian zest'.[37] He does this by shifting the argument under the auspices of the fifth commandment – an effectively diminishing move as not even casuists were conspicuous for excusing adultery. Even Clarkson had contented himself with arguing by implication that Catholic moral theory allows this possibility as it potentially permits incest.[38] There is a contract, argues Mrs Bull, between husband and wife and the wife's duty to obedience is limited by her husband's behaviour. She maintains in marriage her 'Original Right, or rather that indispensable Duty of Cuckoldom . . . . No wife is bound by any Law to which she her self has not consented.'[39] She continues to argue that the very terms husband and wife mean different things in different countries. In some eastern countries husband signifies tyrant, in England the implication is of free and equal government, 'securing to the wife, in certain cases, the liberty of Cuckoldom and the property of Pin-money . . .'.[40] Thus arguments from the terms husband and wife fail to support a doctrine of 'absolute unlimited Chastity'. She continues, 'The general Exhortations to Chastity in Wives, are meant only for Rules in ordinary cases'. They presuppose ability, justice and fidelity in the husband. Thus unconditional fidelity 'could never be supposed by reasonable Men; it seems a reflexion upon the Church to charge her with Doctrines that countenance Oppression.'[41] Throughout this brilliant pastiche there are echoes of the most portentous of seventeenth-century political theories, cribbed as they had been for the trial, but its casuistic nature is underlined by the emphasis on rules applying only to 'ordinary cases', by comment on the entailments of the moral offices of marriage and by the following which I cannot help thinking subverts its own case:

'What is the Cause that *Europe* groans . . . but the tyrannical Custom of a certain Nation, and the scrupulous Nicety of a silly Queen, in not exercising this indispensable Duty of Cuckoldom, whereby the Kingdom might have had an Heir, and a controverted Succession might have been avoided? These are the effects of the narrow Maxims of your Clergy, *That one must not do Evil, that Good may come of it*'.

Thus the cautious casuistic exception to one moral principle, is quite inverted and becomes the moral obligation to ignore another one. As Bower and Erickson note, Burnet had not been forgiven by the High Church for justifying the occasional breach of the law for some good end – as, one might add, he was critical of Cromwell for the same reason.[42] They go on to quote aptly a further worry about the consequences of casuistry. 'We have liv'd too long in those very wrong and absurd Notions of doing *Evil, that Good may come thereby*; to boggle at nothing that will serve the *Ends* of our Party'.[43] It was unfortunate that the justificatory locus for such casuistry was a quotation from St Paul (Romans 3.8). If there is just a hint of subversion in the insinuation that a principled fidelity, 'the scrupulous Nicety of a silly Queen' had helped bring about a European conflagration, Arbuthnot also achieved some distance from Sacheverell's unswerving adherence to the principle of divine succession. As Koster notes, he parodies Sacheverell's high-flown rhetoric, weaving quotations from him and Walpole into Mrs Bull's Vindication,[44] and in the succeeding chapter casts doubt on the adequacy of hardened party doctrines on either side to discriminate between bad and the 'very honest' conduct of some of Mrs Bull's followers. Indeed, it is related to us that husbands, believers all in 'unlimited Chastity and Fidelilty in Wives' went around bullying all wives into repudiation of Mrs Bull's doctrines, and because not all wives would oblige by parting with their 'native Liberty', implacable parties formed, which on all sides bore a nominal rather than real relationship to decent behaviour.[45]

Although undeniably brilliant and powerful, the attack on casuistry, even in *John Bull* is not without discrimination, intimating that Arbuthnot might not be standing quite in the direct no nonsense posture of his High Church Tory protagonist. Indeed, the account of the formation of parties is allegorically at one with Harley's explanation and agrees with the fundamental argument of *Faults on Both Sides* (see above, ch. 2, III). In *The Art of Political*

Lying, as I have argued, the indirection of force is increased and as we may now expect, the parodic treatment of casuistry is similarly less direct and pointed; but a preoccupation with it informs the whole text. The possibility that lying may be permitted *in extremis* that good may come of it is reduced to being the only public obligation, but over which there is no control for the whole unstable relationship between means and ends on which casuistry relied is collapsed. To recall the only quotation from the 'Pseudologia', political lying is defined as the art of convincing the people of 'Salutary Falsehoods, *for some good End.* . . . By *Good*', is not meant any absolute, 'but what appears so to the Artist, which is a sufficient Ground for him to proceed upon'.[46] But how far Arbuthnot himself is diametrically opposed to his Author and Projector is, as I will show unclear. On the surface, in the general context of casuistry, we might expect his opposition to lying to be adamantine, but when placed in more specific counter- point to a recent and very traditional attack on casuistry to which *The Art of Political Lying* alludes, Arbuthnot may, according to his idiom, be invoking a more measured and intermediate response. To explore this possibility, it will first be necessary to outline the specific context of accusations about lying in seventeenth and eighteenth-century England.

NOTES

1   Stephen Toulmin and A. R. Jonsen, *The Abuse of Casuistry: A History of Moral Reasoning*, (Los Angeles: University of California Press, 1988), 'Prologue'.
2   Toulmin and Jonsen, *Ibid.*, for a fine discussion.
3   Toulmin and Jonson, *Ibid.* for the centrality of cases for any fully developed moral theory.
4   On Kant see Kittsheimer, 'Kant and the Casuists', in Edmund Leites, ed., *Conscience and Casuistry in Early Modern Europe*, (Cambridge: University Press, 1988), pp. 180ff.
5   Alexander Ross, 'The Correspondence of John Arbuthnot', Cambridge University unpublished PhD thesis (1956) vol. 2, letter 184, 26 July (1734), pp. 745–6.
6   Gabriel Daniel, *Discourses of Cleander and Eudoxus on the Provincial Letters*, (1694, 1704), p. 101ff.
7   Ross, 'Correspondence', letter 184, 26, July (1734), pp. 745–6.

8   See for example, Plato, *Euthyphro*, trans. H. N. Fowler (Camb. Mass.: Harvard 1947 edn.,) 6D pp. 22, 23. On paradigm cases see also Toulmin and Jonsen, *The Abuse of Casuistry*, e.g. pp. 316–17.

9   Daniel, *Discourses*, pp. 100–101.

10   Daniel, *Ibid.*, Third Discourse at length, esp. p. 127; see also Margaret Sampson,'Liberty and Laxity in Seventeenth-Century English Political Thought', in Leites, *Conscience and Casuistry*, p. 78; Perez Zagorin, *Ways of Lying: Dissimulation, Persecution and Conformity in Early Modern Europe*, (Camb. Mass.: Harvard University Press, 1990), p. 161ff regards the doctrine as the most controversial aspect of casuistry, but see below on equivocation.

11   Meg Lota Brown, *Donne and the Politics of Conscience in Early Modern England*, (Leiden: Brill, 1995) Introduction, ch. 1; Toulmin and Jonsen, *The Abuse of Casuistry*, pp. 250ff.

12   Arbuthnot, *The Laws of Chance*, (1694), pp. vii–viii; *The Usefulness of Mathematical Learning*, (1701), pp. 8–9, 20–22.

13   David Clarkson, *The Practical Divinity of the Papists Discovered to be Destructive of Christianity and Men's Souls*, (1675), esp. chs. 7, 8; Anon., *The Unreasonableness and Impiety of Popery*, (1678) pp. 9–30.

14   Daniel, *Discourses*, at length; Brown, *Donne*, takes up the analogy with equity.

15   Brown, *Donne*, ch. 1; Perez Zagorin, *Ways of Lying*, p. 160. As a descant on this John Toland for example, defined popery as any attempt by priests, protestant included, to think for the laity. See, *An Appeal to Honest People Against Wicked Priests*, (1710), p. 38. Sir Peter Pett had earlier distinguished 'popery' from 'catholicism' but in order to neutralise hostility to catholics, see, Mark Goldie, 'Sir Peter Pett Sceptical Toryism and the Science of Toleration', in W. J. Shiels ed., *Studies in Church History* 21 (1984), p. 261.

16   Margaret Sampson, 'Will You Hear What a Casuist He is? Thomas Hobbes as a director of conscience', *History of Political Thought*, 11 (1990), pp. 721ff.

17   See for example, Daniel, *Discourses*, 3, pp. 101, 108ff.

18   Kittsheimer, 'Kant and the Casuists', p. 187.

19   The point is well made in Brown, *Donne and the Politics of Conscience*; see also Toulmin and Jonsen, *The Abuse of Casuistry*, p. 250ff.

20   The polemical *tour de force* exploiting these potentialities was Blaise Pascal's *Provincial Letters*; but see also at length John Kilcullen, *Sincerity and Truth*, (Oxford: Blackwell, 1988), p. 10ff.; Kittsteiner, and his discussion of Smith's *Theory of Moral Sentiments*, p. 188.

21   Clarkson, *The Practical Divinity of the Papists*, ch. 5.

22   John Wilson, *The Cheats*, (1662, 1671 edn.), Act 5, sc. 4.

23   Daniel, *Discourse*, 3, p. 100.

24   Clarkson, *The Practical Divinity of the Papists* ch. 8, sect. 9, p. 246. See also John Locke, *A Letter Concerning Toleration* (1667), (New York: Prometheus, 1990 edn.), p. 62. Clarkson's views were not extreme.

25   Johann Sommerville, 'The New Art of Lying', in Leites, *Conscience and Casuistry*, pp. 164–5, 176; discussion of these doctrines is taken up again below, ch. 7.

26 Charles de Condren, quoted at length in Daniel, *Discourses*, p. 444.
The unsourced quotation is from *Traité des équivoques* (Paris, 1643)
in *Les Oeuvres compltèes du Charles de Condren* ed. Abbé Pin, (Paris,
1847–8), pp. 415–6. See also Sommerville, 'The New Art of Lying', p.
160. *Traités des équivoques* was commissioned by Cardinal Richlieu.
It is an elegant defence of casuistry from which the author studiously
distanced himself. The Cardinal's reasons for commissioning the piece
may have been precisely because de Condren was not tainted with any
Jesuital associations. Daniel's reliance on de Condren's considerable
moral authority was for the same reason.

27 Because Renaissance and Early modern thought did not reify the
abstraction of 'the individual' does not mean there was no concept of
personhood. It existed as a constellation of offices – one was a parent,
a spouse, a trader, a subject, a citizen, a church member and so forth.
One was also a soul, which however oddly it might seem to us,
encapsulated an office in relation to God. As a soul created by Him,
one could not commit suicide, an issue explored in John Donne's
casuistic treatise *Biathanatos*. The modern individual is the secularised
soul, and it is this secularisation that encouraged the reification of
the individual and the category mistake involved in looking for
such notions of pure individuality and 'self-fashioning' in a world
that conceptually knew better. Some aspects of this argument are
touched on in Conal Condren, 'Liberty of Office and its Defence in
seventeenth-century argument', *History of Political Thought*, forthcoming.

28 Thomas Browne, *Christian Morals* printed with *Religio medici*, ed.
Henry Gardiner, (London, 1845) 1.12, p. 252; John Wilson, *The
Cheats*, Act 5. sc. 4.

29 de Condren, *Traité des équivoques*, p. 413.

30 Arbuthnot, *Reasons Humbly Offered by The Company Exercising The
Trade and Mystery of Upholders*, (1724) in George Aitken, *Life and
Works of John Arbuthnot* (Oxford: Clarendon Press, 1892), p. 381.

31 Sampson, 'Liberty and Laxity', at length.

32 Arbuthnot, *John Bull, John Bull In His senses*, ch. 1, pp. 25ff.

33 Dr Henry Sacheverell, *Speech Upon His Impeachment*, (1710), p. 3.

34 Gilbert Burnet, *The Bishop of Salisbury His Speech in the House of
Lords on the First Article of the Impeachment of Dr Henry Sacheverell*, (1710), pp. 9, 12.

35 Burnet, *Ibid.*, p. 16.

36 Burnet, *Ibid.*, p. 12; cf. John Withers, *A History of Resistance
Practis'd by the Church of England*, (1710), p. 585.

37 Patricia Koster, 'Arbuthnot's Use of Quotation and Parody', *The
Philological Quarterly*, (1969), p. 201.

38 Clarkson, *The Practical Divinity of the Papists*, ch. 9.

39 Arbuthnot, *John Bull in His Senses* ch. 1, p. 25.

40 Arbuthnot, *Ibid.*, p. 26; cf. George Lawson, *Politica sacra et civilis*,
(1660, 1689), ed. Conal Condren, (Cambridge: University Press,
1992), ch. 8, v.

41 Arbuthnot, *Ibid.*, p. 26.

42 Alan Bower and Robert Erickson, 'notes', *The History of John Bull*, (Oxford: Clarendon Press, 1976), p. 157; cf. Gilbert Burnet, *History of His Own Times*, (Oxford: Clarendon Press, 1823 edn.) vol. 1, p. 78.
43 Anon., *Thoughts of an Honest Whig*, (1710), p. 10, cited Brewer and Erickson, *John Bull*, p. 157.
44 Koster,'Arbuthnot's Use of Quotation', p. 207.
45 Koster, *Ibid.*, p. 210; Arbuthnot, *John Bull, John Bull in His Senses*, ch. 2, p. 28.
46 Arbuthnot, *APL*, p. 8; cf. the very similar specification in Daniel, *Discourses*, p. 442ff with the support of de Condren, *Traité des équivoques*. For similar casuistic formulations see Toulmin and Jonsen, *The Abuse of Casuistry*, pp. 203–13, 251–3.

7 Political Lying

'So many parsangs betwixt word and heart'

Robert Burton *Anatomy of Melancholy*, 1, p. 61.

I

Nothing illustrates the case of casuistry as well as dishonesty and it is hardly surprising that the whole seventeenth and early eighteenth centuries have been seen as an age of dissimulation. At the Gun-Powder Plot end of the seventeenth century, the catholic Anthony Copley attacked the Jesuits for their 'Art' in manipulating 'half fac'd Tearmes' and 'whole and demi dublings'[1] Thomas Morton's *A Full Satisfaction*, excoriated against equivocation, 'the notablest Art of lying . . . that ever the prince of darkness did invent'; shortly after Isaac Causabon was to call it a new science; and then Henry Mason elevated the New art of Lying to titular significance.[2] At the Treaty of Utrecht end of the long seventeenth century, Arbuthnot published his proposals for a synthesis of *The Art of Political Lying*. Between these book-ends extended what might fancifully be seen as a pervasive lunatic fog of lies in word and deed, an age of endemic faithlessness, oath-breaking and sundry forms of 'cogging'.[3]

Yet we must be cautious in taking such accusations at face value. Much the same has been said often enough of our own era. The Reformation was a fragmentation of Christendom accompanied by strong imperatives to orthodoxy. As Zagorin notes, wherever people were likely to be compelled to come into a local church, they might be suspected of dissimulation and outright lying and so on all sides the contentious issue of dishonesty and insincerity was explored on the basis of established authorities.[4] The notion that the world was entering a new age was, however, no neutral description given the strongly prejudicial theological associations of 'new-ness' and novelty. The accusation is intrinsic to the descriptive claim. Henry Mason was insistent that Jesuit doctrines of reservation and equivocation were distinctive and that Catholicism as such was by no means similarly tainted, the art being 'unheard of before these latter dayes' and most developed by English Jesuits.[5]

Arbuthnot's echo of these old accusations in his reference to unspecified new discoveries in the art of lying together with reference to his Author's amassing instances from before as well as after the Revolution require a context broader and deeper than recent Whig propaganda. If so, his satire stands as an ironic abridgement of the century's dominant combative theme.

II

On the surface the moral status of lying seemed clear enough in early modern Europe. There is an unequivocal eighth commandment: 'Thou shalt not bear false witness' and Arbuthnot was not alone in exploiting Satan's status as the father of lies and so making lying a monstrous sin. As Sommerville correctly points out, however, The Bible is itself quite equivocal on the blanket prohibition against lying, as was Roman rhetorical theory and early Christianity.[6] Cicero adumbrated upon the associated tactics of obfuscation and prevarication as arising from the rhetor's need to be able to argue any case *ad utramque partem*. As a consequence he came perilously close to compromising the rhetor's august *officium*.[7] For all his similar emphasis on probity and upon rhetoric's serving the truth, Quintilian allowed lying in a good cause and even defended his position in a way suggestive of later definitions of dishonesty. To the accusation that rhetoric is a matter of lying he claimed that rhetoric served the truth and only made use of lying, the rhetor might well deceive others but not himself, just as the good general never believed his own deceptions. Again, and in a way strikingly suggestive of Arbuthnot's politician who also is enjoined not to believe his own lies, if rhetoric is an art, the rhetor is an artist and he deceives the observer by his technical skills, but never himself.[8] The unqualified prohibition on lying was to be most authoritatively elaborated by Augustine. In *De civitate Dei* he would insist that there was honour even among thieves, the point being to stress that society required a degree of trust, even if it was itself untrustworthy and illegitimate. His most authoritative accounts of this issue, however, *De mendacio* (c.395) and *Contra mendicium*, (c.420) were written in immediate dispute with the Priscillianists and with St Jerome. Less directly he levelled his sights against the whole tradition of rhetoric in which he had been educated and thus against Quintilian's slippery justifications in the name of truth.

Against the Priscillianists, he argued that honesty had to be maintained among all peoples, there could not be one rule for Christians another for their dealings with others. Not all lies were equally wicked, to be sure, and the greater good contingent on lying, the less serious the sin, but sin it always remained.[9] Because of the public significance of rhetoric, it was a sin endemic to the world in which he was born. Late Rome, it seems was yet another age of lying. The rhetorician is trained to lie and Christianity must stand against such institutionalized evil if it is to seek the way, truth and light.[10] Against Jerome, he maintained there could be no lies for some good end.[11] Augustine's views were later endorsed by the authority of Aquinas,[12] and then by Luther and by Calvin against what he deemed Nicodemism.[13] Of all the condemnations of lying, *De mendacio* may be seen as standing distantly behind *The Art of Political Lying*, for as I have suggested, Arbuthnot is providing a parody of rhetoric and its claims (see above, ch. 5, II–III). It is rhetoric reduced to its most Augustinian, or perhaps Pascalian dimension by one who, *prima facie* might seem unbendingly Augustinian in his condemnation of Lucifer's progeny.

Given the principle which seems a little like Pope's 'misty' generality, what, in particular cases is being condemned? There is something deceptively tidy about lying. It seems part of a simple binary division with truth, something appealing in a post-Ramist world, suggesting that understanding the nature of a lie should be straightforward. It is a view that is still with us in as much as it is lies, not truth telling that require explanation.[14] Arbuthnot attests to such expectations by having the Author of the 'Pseudologia' call lying an art because to tell the truth requires no art.[15] Indeed, conventionally, defining a lie has not seemed to pose many problems. From Augustine to Montaigne, to Bok and Barnes, it has been predominantly understood in one of two related ways. As a known discrepancy between what one understands to be the case and what one says, lying is condemned as an act of hypocrisy and insincerity, it is a dark fissure between heart and word, an untruth knowing told.[16] Lying could also be understood as an untruth expressed for the purposes of deception. This restriction on the notion is a development of the acceptance that some untruths might be worse than others.[17] It was this species of definition, in contrast to the Augustinian emphasis on a formal discrepancy between statement and understanding which was to be developed by the casuists and upon which the definition in *The Art of Political Lying* is a descant. Each understanding, however, involved an appeal to the inner workings of the accused, and so it is with the criteria for the use of the definition in terms either of sincerity or intentionality that problems begin to emerge.[18] In a more practical way, neither species of definition permitted uniform condemnation of any designated lie, so the very power of the accusation could be problematic. Cases which might seem marginal in terms of one feature of lying begin to confuse the process of applying the principle. Augustine himself laid the seeds for this, arguing that some lies were worse than others and by trying to implicate the whole of rhetorical education in lying. In arguments about the nature of lying entangled as they are with specific accusations about dishonesty, we see clear examples of the contested character of language and specifically a case study in a semantic war of attrition over the range and interrelationships of salient terms. Where some writers, following Augustine, sought to give the term 'lie' a maximal extension, sometimes eliding it with putative consequences, others tried to restrict its scope, by insisting that a lie entailed an intention to deceive. The resultant increase in the accusation's intension required the introduction of more benign terms for what was thus excluded.[19] There was a good deal at stake in distinguishing, for example, a lie from a fiction or refusing to do so. To pursue such issues systematically here would be distracting.[20] What can be said of the whole early modern period is that the lie, with paradigmatic cases of malicious false witness and oaths sworn in bad faith at its centre, was itself the very model of deceit, at the still black centre of a fuzzy-edged family of sins, faithlessness, cheating, dissimulation, hypocrisy, fraud, perjury and rebellion.[21] Each undermined the sense of trust on which society depended and so had serious consequences. Formally opposed to truth, then, lying could be construed as threatening society itself; it was certainly the necessary condition for conspiracy and plotting.[22] As Morton strikingly put it, lying created a sort of Gyges ring invisible to Protestants, so enabling men to plot as they will.[23] So tightly bound were moral sin and political consequence that the notion of a society of liars seemed, as it had for Augustine, a contradiction in terms, a possibility which Arbuthnot tries to make imaginatively plausible.

All this made the accusation of lying even more potent than it is today often shorn as it is of its cosmic associations and penumbra of social solvents; but it was also, as it remains, even in its crudest forms a most economical accusation. Swift remarked that a lie believed for only half an hour had done its work, an accusation of lying believed for half that time is as effective. The accusation, 'You lie, Sir', by making one explicit point , 'x is untrue', at the same time attacks the speaker's intentions and integrity. It entails an element of the *ad hominem*, and to the extent that it is an accusation, it affirms or reaffirms the moral integrity of the (shocked, certainly righteous) accuser. The lie contains a world of rhetorical partition and such economy has often proved irresistible and made defence difficult and convoluted – an ideal challenge for the casuistic mind.

III

With stark honesty Machiavelli had re-aligned debate away from what might now seem to be the rehearsal of post-Augustinian pieties. To be sure, Machiavelli does not come like a bolt from the blue, there had been a distinctly casuistic strain to medieval legal reflection on the status and responsibilities of rulers.[24] Cicero and Quintilian were entrenched in Renaissance education, and their casuistry on behalf of rhetoricians could hardly have passed unnoticed. Nevertheless, Machiavelli accepted in the most peremptory fashion that princes gained and lost power by force and fraud; and a whole air of dissimulation prevades his advice. There is a difference between appearance and reality; seeming to have conventional virtues is more important than exercising them; it is necessary to know when to break faith, to play the fox as well as the lion.[25] In short, given the world in which a prince must perforce live, one of uncertain trust infused with deceit, he must be prepared to lie, cheat and maintain an air of hypocrisy if he is to survive. There is no pretence that this is really good behaviour or that it is necessary for the public good, or is only an aspect of the onerous responsibilities of rule: the prince must simply know how not to be good. This is offering casuistic advice in its most simple exceptional form and the form most exposed to a morally absolutist counter-blast. Necessity, emergency allows all forms of dishonesty. Only the Prince can gauge that necessity, a large part of his *virtù* lies in doing so and grasping the occasion. It was this whole aspect of *The Prince* which so troubled Machiavelli's critics as both immoral, indeed devilish and counter-productive, because for conduct to be based

on such advice destroyed or further eroded the trust that made
society, even among princes possible.[26] It was certainly an inadequate
defence against accusations of lying and it is little wonder that for
a writer like Justus Lipsius who was much impressed by Machiavelli, the exposed acceptance of a necessity to lie is re-described as a
matter of 'mixed prudence', but now justified in terms of the public
good.[27] Similarly, the young Thomas Hobbes, (?) if he is alluding
to Machiavelli, distances himself from a doctrine of necessary
dishonesty in politics, by remarking of the Emperor Augustus that
'in those times' dissimulation was 'held an inseparable accident of
a politic Prince'.[28]

As I have noted, Arbuthnot knew Machiavelli's works directly
and in another context applauded the Florentine and it is surely this
notorious aspect of *The Prince* of which *The Art of Political Lying*
is in part a reduction. It posits a world in which lying has become
the only necessity, where the casuistic exception has become the rule
and in which each participant in the system is a sort of prince,
drawing no distinction between his own and the general interest and
proceeding accordingly. That Arbuthnot's political system can seem
to work as a society is because it is a world without violence. To
re-iterate, even his predominant image of rhetoric, the life-force of
his political system, is one of creative negotiability more than open
verbal war.

Much as the Machiavelli of *The Prince* might be reviled – to do
so was after all an expression of one's own integrity, his views had
to be taken seriously in the wake of the Reformation which so
disrupted Christendom. The dilemmas of the courtly world were
spread outwards and downwards. When true religion (God and his
honest servants) were being persecuted was it permissible to lie for
it? That is, could cases of lying and deceit be re-classified as cases
of a higher loyalty and if so, who was to judge and by what criteria?
If not, how close could one come to lying and not sup with the
Devil? It is here that we shift from the exceptional casuistry of
Machiavelli, to the more typical extensive casuistry of Reformation
Europe; the morality of the long spoon. From lying out of una-
dorned self-interested *necessità* we shift to propagating, as Arbuthnot put it, in more Lipsian terms, salutary falsehoods for some
good end, matters of the mitigation or the dissipation of the
accusation itself through a re-description of the act as a reservation
or equivocation in the name of charity or justice or some generally
useful consequence.

IV

Historically, just as it had been possible to move from a paradigmatic case of malicious false witness, to a very extensive and
accommodating notion of lying around which orbited a whole
constellation of satellite sins, so a defensive semantics of lying
began to be deployed. As Montaigne had remarked, there may only
be one way of hitting a target but there are many ways of missing
it. Such misdirection might lead to radical confusion, as Sir Thomas
Browne put it, 'some truths might seem almost falsehoods, and
some falsehoods almost truth'.[29] Thus one finds that the accusation
of lying was softened, or dissipated by employment of a rich set of
associated terms used to re-describe what is said about the politico-
moral world. Erasmus had noted that what Augustine called a lie,
Jerome considered only dissimulation and although himself not in
favour of lying could see circumstances where a good man might
need to dissimulate.[30] But gradually the re-descriptive vocabulary
was elaborated from such tentative beginnings. A speaker accused
of lying might be said to be equivocating, prevaricating, evading,
exaggerating, being discreet making a reservation giving us the
truth by halves, or disguising the truth.[31] As I have already quoted
de Condren, (ch. 6, IV) the business concealing the truth under
figures, through *paradiastole* was permissible,[32] but the very idea
that one is concealing the truth by a range of contingent means, is
to have re-described what others might call lying. As Henry Mason
argued, the very notion of equivocation is a shame- faced avoidance
of the correct term, lie.[33] The language of lying was full of 'leaps',
turnings', and 'juggling tricks'; the 'half fac'd Tearmes' and 'demi
dublings' in which Copley had seen so much art.[34] As Clarkson
later put it prejudicially, it's all still lying, one may lie plainly, or
secretly.[35]

And there was an impasse. Neither side could appeal to principles
to settle what was what, for all agreed that lying was wrong; and
nor could there be a repair to cases, for these had first to be
described.[36] So it remained throughout the seventeenth and early
eighteenth centuries, one man's secret lie was another's fiction,
reservation or undercover truth. To one set of ears the speaker may
only be withholding information using an ambiguous term, offering
half a truth, or saying nothing, to another this may effectively have
been lying. This last possibility is a crucially complicating factor in
the rhetorics of political truth, for so much in politics is a matter

of inference, innuendo and implication.[37] Thus even literal truths can be uttered with the intention of deceiving.[38]

From its earliest years, this question of contextual truth was seized upon in Jesuit casuistry. A statement or half statement taken in isolation may seem to be true or false, one requires a fuller context to know if it really is. Thus it was that Navarrus, (b. 1492) building on the notion of mixed modes of speech, (fully explicit, implicit and silent or, as is the norm, partially explicit), developed his notion of *amphibologia*.[39] This was a co-option of a traditional rhetorical notion of amphiboly, from the Greek expression meaning to cast a net. Traditional amphiboly was an argument that exploited ambiguity in language so that different parts of an audience could be gathered in agreement with the speaker. For Navarrus, *amphibologia* involved a mental distance from one's language irrespective of its inherent potential for amphiboly in the old sense. It was not lying, it was reservation or equivocation, resulting in a mixed or mental proposition. In his unpublished *A Treatise of Equivocation* Henry Garret accepted such arguments so that the Catholic Church in England might survive persecution. He was attacked by Catholics and Protestants and his case defended and extended by Robert Persons.[40] In agreement that lying is wrong, each confronts the hard issue of what a man should do if, rightly sheltering a priest from persecution, he is questioned by the persecutors. Each advises that it is permissible to deny hiding the priest. To paraphrase: you may say no to the man's face if there is no softer way of suggesting the negative, reserving the fuller statement 'at least not for you to find' for God. Thus one does not bear false witness to God and a good end is served. It was exactly this line of argument which so worried critics of casuistry and which was seen as being at the heart of the new art of lying. To isolate this, Thomas Morton and Henry Mason drew a distinction between open equivocal language, with all its potential for amphiboly in the traditional sense, logical equivocation which grew from the ambiguities of language, and mental reservation, a 'new bred Hydra and uglie Monster'.[41] Once sanctioned, such reservation had the effect of destroying the reliability of public discourse, replacing it with a solipsistic private world. John Barnes, not to be confused with John Barnes, seems to have gone further, mental reservation was the invention of the devil and worse than lying.[42] The Jansenist critics of the Jesuits were no less worried and some Protestants were not slow to taint Catholicism with the whole tar-brush of Jesuitry – a trifle unfair as not all Jesuits were casuists and quite a few Protestants were.[43] On both sides of the Reformation divide the fear was that to allow lying or the benign re- specification of it led to the evaporation of almost any sin. As the Jansenists in particular realised, sin itself, thé most abstract of negative principles could be distinguished virtually into nothingness.[44]

V

What needs emphasizing here is the well recognized contingency of the descriptive process. Unless some stable independent criteria could be found to limit re-description, that is to contain *paradias-tole*, the very distinction between truths and lies would be eroded.[45] This is a realization that seems to date at least from the dialogues of Plato. In the *Republic*, when charting the collapse of oligarchy into democracy, Socrates lists the sorts of vices that may be dramatically transformed (*metabāllei*) into virtues, insolence becomes good breeding, prodigality magnificence and liberty licence. Plato, of course, assumed a proper standard, it was one of the functions of the theory of forms to provide such a court of appeal standing beyond the vagaries of language.[46] Thereafter in dealing with the problems of moral re-description there are two distinct types of *paradiastole*. Mostly, there is the limited kind for which there is taken to exist a stable criterion (such as a form, truth or raw hard reality) to govern re-description. For Henry Mason equivocation in the Jesuit sense was but a new name for an old thing-the right name is lie.[47] More generally, wrote one of Arbuthnot's contemporaries, we have been taught to call things by false names.[48] The problem is presented as involving a simple appeal to the facts against which to measure linguistic deviation. Indeed, the choice of the word 'thing' to stand in opposition to 'words', seems to mark such an appeal to a commonly shared solid reality. Words are wayward, things are real and should have only the proper labels which all of us should know. As we know from the Author's catoptric simile for the soul, it is the function the plano surface to present *things* as they are, the cylindrical to present them as they are not. In an ideal Laputian world even true labels would not be needed and there would be no problem of redescription at all. *Paradiastole*, however, is controlled only all the time there is agreement on the criteria for language use, all the time there is a common world of shared concepts or 'things'. If there are no agreed standards, if say, there are no known Platonic forms or other universal truths, if the 'things' are not accessible without some variable predication process, a more radical form of *paradiastole* becomes possible. This was put to the foreground by Montaigne's essay 'On Cannibals', which concluded with no arbitration as to who are the true barbarians, the Indians or the Europeans.[49] Again John Earle explored briefly the character of the sceptic, one who believes we take our ideas as the air we breathe and beyond this there is nothing enabling us to judge between such ideas.[50] Such scepticism thus confounds the objective use of words like 'lie' and makes the accusation that we have been made to call things by false names a case of *petitio principi.*

Henry Mason had explicated just how the art of equivocation undermined society because it undermined the stability of language. Equivocations in language, created by forms of ambiguity and amphiboly are one thing, for they are at least apparent to people with sufficient linguistic competence,[51] but half statements qualified or contradicted by mental reservation, destroy all common criteria for the use of language, we can say what we like and nothing can be known to signify.[52] This, of course, is a theme also elaborated by Hobbes in whose hypothesized natural condition there is no agreed upon signification for words, and whose sovereign, as a corollary, must have the right to arbitrate on matters of public meaning and doctrine.[53] As I have intimated *The Art of Political Lying* may be seen as alluding to a Hobbesian state of nature and this possibility is strengthened in as much as the natural condition is a state of radical *paradiastole* which Hobbes illustrates with reference to the words good and evil, terms which subsume understandings of truth and falsehood. In the natural condition each calls good what he likes, evil that which he dislikes or fears. There is no authoritative voice of arbitration, no criterion to be univocally applied, and so there is no stable public language. For Hobbes it is this intolerable condition which necessitates and explains the creation of the Leviathan who determines public truth, public good.[54] Swift would provide ironic allusion to this problem in its most nominalistic and materialist form by having men walk around weighed down with objects rather than rely on the hopelessly equivocal and contestable words of reference.[55]

Arbuthnot's world is Hobbes's natural condition turned very much on its head. Or, put another way, it is a thought experiment with just the kind of social scenario that Mason feared must arise from the acceptance of equivocation. The political system is less a state of war than a state of fraud. The art as presented by the 'Pseudologia' is the art of appropriately imaginative redescription where the only limitation on its most radical form is the probability of what will be believed. This is indeed, Cicero's notion of argument as 'a plausible invention (*inventum*) to gain conviction'.[56] The taxonomy of lies reported from the 'Pseudologia' is a set of rules for *paradiastole*.[57] The guidance is all about how to describe and re-describe the world so that accusations, rationalizations and a whole range of injunctions become plausible. It is in the technical sense a matter of *inventio*. The motivations specified for creating translatory, detractory and additory lies make it clear just how arbitrary the whole political vocabulary must become; the artists lie in the name of justice and the public good, but these terms can only mean what they want them to. The result as I have already suggested (ch. 5) is to invert Shaftsbury's optimistic notion of common sense, the socialized aquisition of truth for the common good, out of respect for the rights of humanity. In the process Arbuthnot elides the distinction between the imaginative, constantly re-descriptive language of politics with a world beyond words. This is not just a logical consequence of the system as we have it described, it is manifested in the terms chosen to designate the rules the artists should follow in elaborating plausible lies. Sometimes the Projector writes of what can prudently be said or put about concerning political figures; but he also writes as if the artist is directly creating the subject. Rhetorical *inventio* shifts into the inventiveness of the painter. The irreligious man is not to be *made* too devoted but 'you may with Safety make him sit out public Prayers with Decency'; 'One would not make . . . an Atheist support the Church'.[58] Many recent lies have not remained within the bounds of probability and have proved 'abortive or short liv'd'.[59] The metaphor of creating life for lies is striking, as is the image of the reified lie, 'the *Raw-head and Bloodybones*' being brought out as a terrifying 'Object' to frighten people.[60]

The cumulative impression is of making and saying being much the same. The artist is almost a maker of physical reality, that familiar poet, painting blind pictures, the artist of silent poems.[61] Above all, the evocation is of a rhetorician of Gorgian reconstructive capacities. To recall Peacham's more immediate hyperbole, the rhetorician in persuasiveness is next to God, which is to

come perilously close to being divinely creative. In this context of
associations, we have a full reductio of the argument by Quintilian,
that the rhetor is a true artist, like the painter creating pictures so
realistic they deceive.[62] As a corollary, Quintilian's distinction
between rhetorical and artistic knowledge becomes a casualty; that
is, knowledge for action (rhetoric) and knowledge culminating in a
creation amount to much the same thing once the unfortunate
Quintilianesque metaphor of the rhetorician as artist lying for
reified truth is taken with due Arbuthnotian *gravitas*.[63] This is not
to suggest a single locus for the redescriptive play with words and
physical creation. In Scots, the word *maker* could signify a poet;
some thing directly analogous was the case in Greek and this would
hardly have been lost on Arbuthnot.[64]

Reputation, standing, a person's credit in society, was vital for
social and trading relations, especially in a world in which hard
cash was often scarce, but credit and reputation, the essence of the
social self, was a function of what was believed of and said by
others.[65] Arbuthnot's artists are makers and breakers of reputation
and credit through words. Indeed, in the account of Chapter Five
of the 'Pseudologia' devoted to the main strategies for making lies
believable, all verbs of discourse (such as state, say, tell) presupposing an independent identity to be talked about are replaced by verbs
of action assuming direct powers of creation and control, 'make',
'allow', 'bring'.[66] Not surprisingly, the question of individual identity was much in the air in Arbuthnot's world; Locke had explored
issues concerning individuality in terms of corporeal extension and
personality and these grew out of well established problems of
distinguishing a self from a series of social roles.[67] The Scriblerians,
like satirists throughout the century,[68] were to have much convoluted fun with the twists and turns of identity in parody of Locke.
More immediately we may detect a further association of Arbuthnot's informing metaphors of money (see above, ch. 5, IV); there is
an intimation of a broadly Tory distinction between landed and
monetary wealth which coupled the latter with an appropriate
personality type – shifty, self-interested and insipiently fraudulent.
If as Lawrence Klein remarks, the eighteenth century politics of
politeness made a virtue of plasticity, Arbuthnot makes mutually
maleable identities the essence of a political system.[69] Arbuthnot's
own choice of language, then, in *The Art of Political Lying*,
provides a variation on this theme of the stability of personal
identity. If political reality is nothing but a rhetorical system for

redescriptive creativity, its participants have no more identity than the velocity and stability of lies told about them. And unless we take Arbuthnot literally, *The Art of Political Lying* is itself an act of *paradiastole*, where all political discourse and the institutions of politics are re-described in terms of an elaborate and endless glass-bead game of lies. One of the main problems of interpretation inheres in the uncertainty of reconstructing what he is reserving to himself despite the words of the Projector. Thus we have the full slippery significance of the Projector's enthusiasm for the wealth of modern *inventions* in lying of which we are informed at the outset. Arbuthnot is playing with the ambiguities of *inventio*, imaginative creation and fabrication, so fashioning a notion of lying sufficiently central and extensive as to leave nothing in political discourse beyond its range.

What I have been suggesting illustrates the commonplace that areas of a culture often develop mighty coagulations of terms and expressions around forms of experience of intense importance within them. One purpose is for increased precision. This is what both Hobbes and immediately before Arbuthnot, Locke had presupposed. Words, Locke argued, had to be sufficient to the ideas that needed to be expressed, to be serviceable it was necessary that there be no dissonance between idea, word and hearer's understanding of the two.[70] Such an unbroken continuum, furnished communication with integrity, even if it was in error. But another reason for semantic diversity is for increased descriptive flexibility in order to help avoid being tied down disadvantageously. The language of politics can make, indeed has made the lie as difficult to get hold of as it is so central, because it is so central. As Barnes remarks, in many societies, those seen to be in power are most likely to be subject to accusations of lying and so in need of mechanisms of defence.[71] It is not just that it is unhelpful to classify all statements simply according to the dictates of a truth table; it is that in political discourse where truth claims and mendacity accusations are the stuff of life, vocabularies are developed like immune systems. *The Art of Political Lying* may be seen as an ironic response to this phenomenon. From what is related of the 'Pseudologia', the work itself develops an elaborate set of distinctions which do indeed give precision and at the same time it outlines the rules of claim and counter-claim, accusation and appropriate defence to allow the players in the game to avoid being terminally and terminologically discredited.

Overall, where the casuists could by increasing semantic precision shrink serious sin to a pin-head on which angels would have trouble alighting; their opponents, by reverse feats of semantic extension, provided a veritable *chaise longue* of sin on which anyone might be strapped. Neither conformed to Lockean hopes for the confluence of word and idea. Where Arbuthnot chose to sit in 1712 is not as obvious as it might seem in the light of his much later Pascalian poetic effusions. There can be little doubt that the Jansenists and their Protestant allies provided a species of critique which weighed heavily with Arbuthnot; yet the very strategy he adopts in *The Art of Political Lying* of suggestion and innuendo, is a form of casuistic reservation and equivocation, entangling the reader in his own 'Salutary Falsehoods *for some good End*'.

As for the Lockean notion of the serviceability of words, that is, the communicative continuum from belief or idea in the speaker's mind to the reception of the speaker's words, *The Art of Political Lying* simply distorts the criterion for sound word use. Rather than exciting 'in the hearer exactly the same idea they stand for in the mind of the speaker',[72] they must always excite something different, lest the artist starts believing his own lies. The continuum must be to some degree distorted. As I will suggest in the following chapter, a substantial part of that '*good End*' for Arbuthnot was to force the reader to confront the difficulties involved in the slippery relationships between truth and falsehood, by suggesting opposing patterns of proposition. And we can now suspect that the very force of the Author's injunction that the artists must not believe their own lies is problematic. Conventionally such self-delusion could only exacerbate the sin of lying. Yet Pierre Bayle had already published a subtle argument derived from Augustinian premises that led to a different conclusion. Bayle shared what would be the similarly conventional Lockean view that there should be an unbroken continuum between belief and the reception of words. He also insisted like Arbuthnot that in moral and political matters knowledge was largely a matter of probabilities, not deductive certainties and simply demonstrable errors. But he argued further that people could behave rightly and with integrity on the basis of genuinely held false beliefs. Truth and error was one thing, the distinction between truth and dishonesty required that words and actions be discrepant with belief.[73] In the context of this variation on familiar themes, the Author's injunction for politicians not to believe their own lies carries a satiric force distinct from additional condemnation of what is already bad enough. Among those who merely spread lies, self-delusion may have some merit, but in the midst of the political creators it threatens the very existence of the elaborately crafted system of political dishonesty. Truth may still be absent, and error might abound, that is always likely in a world governed by probabilities, but errors invented in good faith infect the system with integrity at its heart. If politicians believe their own lies, the Author's work is invalidated.

VI

Moving from the most general context of casuistic controversy and Arbuthnot's play with the vocabulary of lying to the more immediate world behind *The Art of Political Lying*, one can at least say that the pattern of discourse provides an index of the fragility of trust, a hall-mark of the rhetorics of party division born of and sustained by that fragility. But whether the early eighteenth century really was so pocked with cheating, is another matter.[74]

The repeal of the Licensing Act in 1695 in rough conjunction with the Triennial Act, may well have unleashed extraordinary pamphleteering energies, but the greater freedom to print is more deviously related to the prevalence of accusations of lying. Such claims and counter-claims had after all run through the seventeenth century. Increasingly from the mid-seventeenth century, London had supported a well organized system of printing and distribution. As had long been recognized, London was politically so central to the stability of the country as a whole, that to be able to control or shape London's public discourse and patterns of political perception was seen as vital. Hence the alliance between politics and letters which certainly pre-dates the reign of Queen Anne. But there could easily be too much of a good thing. Put another way, the print controversies of the early eighteenth century may have manifested a version of the prisoner's dilemma: if a few distribute a little news it might get adequately circulated, be informative and be critically appraised or at least prove, in the absence of much else, plausible. But with a plethora of people of diverse perspectives with differently presented bits of information trying to participate in such a process, the effectiveness of distribution is diminished and debate is undermined; almost anything might seem a half-truth, anything some form of dishonesty. It may have been the very flow of informational snippets, designed less to inform than muster and defend, that helped intensify the heat, and give the expansive impression that people were living in times of endemic cant (see above, ch. 2, II). Then as now the apparent ubiquity of lying may be a function of information not disproportionate impropriety. There is, however, a more specific reason which helps explain the prevalence of accusations of bad faith. In the years approaching the Treaty of Utrecht, foreign policy was a domestic issue and we may expect increased fears of political dishonesty when the *arcana imperii* of diplomacy and secret negotiation which nearly always appear to involve a form of institutionalized reservation and equivocation are prime candidates for public scrutiny. A similar situation can be seen in the period from 1677 when it was Charles II's feared negotiations with France that became the catalyst for a serious breakdown of trust within the political arena and a consequent outbreak of conspiratorial and covinous indictments.[75]

To catalogue such accusations at length in the years immediately before the printing of *The Art of Political Lying* would be to tell a tale as tedious as it would be long but some of those which seem to be picked up or alluded to in the tract should be noted. Davenant's character Tom Double, encodes dishonesty in his name and makes much of the importance of spreading lies, no matter how gross, and according to season throughout the kingdom.[76] This attests to the recognition that the distribution of print and communication by word of mouth was the means by which a world of public opinion was made part of a political system. It is this, as it were, Habermasian theme that Arbuthnot institutionalizes and codifies so clearly with his references to the different velocities of lies and the vital role played by dissemination as well as creation. Dunton, well attuned to the atmosphere of accusation devotes several of his paradoxes to the theme. Apart from a versified paradox in praise of a lie he has a consoling essay in praise of slander and an argument that it is better to be chaste and honest, a paradox in the context of there being so much dissimulation in the public realm.[77] Swift's *Examiner*, alluded to the many new improvements in the art.[78] The Tory paper *The Plain Dealer*, printed by John Morphew, emblazons a colloquial honesty in its title and frequently resorts to accusations of lying. It demands censure for those who publish falsehoods, it purports to set down plain rules for controversy; it complains of lies being spread abroad to discredit the present ministry and of Whigs distinguishing truth into malicious scandal and so forth.[79]

What is perhaps more important than the ubiquity of this species of accusation is its conceptual extensiveness and its semantic richness. Nothing illustrates better that after a century of casuistic debates the classification of a lie is a porous one and because of this easy to use and elabórate. *The Plain Dealer*'s advertisement for the Six volume History of Whig lies gives a good flavour of the semantic richness of the accusation; referring to conjectures, surmises, long-lived stories, insinuations, rumours, dogmatic assertions and short reports.[80] A few months before this, Addison made clear how lying could so easily extend from directly dishonest denial to a complete education in political conduct.[81] Jesuits like Daniel might cry in vain for a little discrimination, and attempt to restrict the semantic range of the term lie at least by contrast with a notion of fiction, or half truth but all to no avail.[82] In what is tantalizingly suggestive of the political scope of *The Art of Political Lying*, rather than its precise form (see above ch. 3, I), Addison announces a new French Academy for aspiring politicians in which one Master is to instruct in 'State Leger-demain'; another 'Artist is to teach them how to nod judiciously, shrug up their shoulders in a dubious case (and to teach) the whole Practice of *Political Grimace*.' A 'Language Master' is to instruct 'young Machiavels' in diplomacy. Even the most innocent of questions must be avoided or answered misleadingly: if asked the time one should answer indirectly, or 'turn off the Question'. If asked for the news one should deny having read it. An elevation of the 'left Shoulder' will stand in lieu of an explanation. A further Professor, is likely (predictably) to be a Jesuit, chosen to teach probable Doctrines, mental Reservations . . . how to distinguish the Spirit and the Letter . . . the Art of finding Flaws, Loop-holes, and Evasions . . .' and of proving that contradictions may be equally true and valid. Another master is to give a final polish in what is shortly to be the Arbuthnotian promissory lie, 'by furnishing them with Bows and Inclinations of all Sizes . . .' It is, Addison concludes, all an exercise in 'French Truth'.[83]

VII

In such a broad context of satiric play with lying, it is not surprising to find an honesty/dishonesty *topos* central to all of Arbuthnot's satires of 1712 and that in a sense *The Art of Political Lying* formalizes what is often implicit or inchoate in the *John Bull* satires.

The terms of this *topos* are varied. Apart from honesty, truth and lying we hear of cheating, plain-dealing, leger de main, tricks, infidelity, breach of trust, and perjury. The climax of the whole sequence involves Lewis Baboon turning 'honest', which presupposes previous deceit, but also illustrates how accommodating the accusation of lying could be. Turning honest is settling accounts in a way satisfactory to John. In fact, apart from John Bull's mother, The Church of England, there is hardly a character who is not seen as in some degree, dishonest. The First Mrs Bull is presented as a believer in the rights of Cuckoldom. She is an amalgam of Davenant's principled Whigs, those who believe it right in principle to break faith by deposing a legitimate monarch, and 'modern' interest-driven Whigs, those who serve the faction out of convenient self-interest.[84] As Patricia Carstens points out, Tories had been apt to accuse Whigs of casuistry, and in Mrs Bull's Vindication we have a straight-faced set of allusions to casuistic principles of argument.[85] Chastity applies only in 'ordinary cases' and infidelity provides the necessary means of perpetuating families.[86] The document was found by John in her cabinet after her death, and shocked him, a reiteration of the lack of trust between them.[87] Nick Frog is the quintessence of self-serving dishonesty, who lies constantly to John; Hocus is not much better, his hocus-pocus being designed to deceive, though the interest that motivates his lies is different. Even good Sir Roger Bold (Robert Harley) is implicated in the breach of faith that leads Jack (Presbyterianism) to his death, despite Jack's friends' disingenuous promises to rescue him.[88] John himself is 'in the main an honest plain dealing fellow'.[89] Given the slippery extensiveness of the lying *topos*, 'in the main' is as much as we can plausibly hope for. Additionally, the mechanisms by which the villains try to dupe John are largely exercises in stage Machiavellian duplicity (Nick Frog shines here), protecting self-interest in the name of necessity above everything else, or they are parodies of casuistic, even mechanically casuistic mitigation. Don Dismallo even offers John a selection of argumentative strategies to listen to, like offering a choice of weapons in a duel. At the end, when Lewis Baboon does indeed turn honest, John can hardly believe him enquiring suspiciously if the '*Equivocals* and *Mentals* 'have been set aside'.[90] In short, the action of the whole set of pamphlets is jogged along by various forms of dishonesty, the attempts to dupe John, and the revelation of truth and interest, *veritas* and *utilitas*, coinciding only at the end in a triumph over casuistry. As the whole story

is an allegory of recent British and European politics, we have the
beginnings of a world that leaves little room for anything but forms
of lying, suggesting the art and artifice of a world operating
according to its own arcane rules.

VIII

Leaving aside Addison's French Political Academy with its appropriately complete education for such a realm, an immediate catalyst
for Arbuthnot's elegant assertion of this sardonic vision may well
have been a specific sermon preached before Queen Anne. It is
alluded to by the Projector in one of his pointed hesitations in
accepting the views of the 'Pseudologia'. The sermon, along with
unspecified passages from the Bible creates, we are told, 'seeming
difficulties' for chapter Three of the 'Pseudologia', which adum-
brates rights to truth and the lawfulness of lying.[91] As the sermon
was delivered by William Reeves whilst Arbuthnot was attending
The Queen at Windsor it is likely that he heard it; but irrespective
of this, she had ordered its immediate publication and it was in the
public domain before *The Art of Political Lying*.[92] Patricia Carstens
has shrewdly suggested that Arbuthnot's essay is something of a
parody of Reeves' sermon.[93] Up to a point this seems right, but as
I shall suggest we cannot infer a straightforward parodic endorsement of everything Reeves urges. A simple Reevesian hostility to all
lies is rendered implausible by the nature of *The Art of Political
Lying*, by Arbuthnot's previous political writings and his continuing bouts of political insouciance.

Carstens points to organizational similarities between the sermon
and *The Art of Political Lying* and claims that Arbuthnot's additory, and detractory lies are transformations of Reeves' division of
malicious lies into slander and defamation.[94] The one is a matter of
'loading' people with vices of which they are innocent; the other
involves 'clipping' their virtues – in one sense an appropriately
monetary image for Arbuthnot's purposes.[95] We must be cautious,
however, in claiming a direct and derivative relationship. This is not
just because each man was using a common stock of language and
sentiments (Reeves' understanding of detraction and slander were
public property); but also because any derivation, or inspiration
from Reeves, is augmented, inverted, or exaggerated by Arbuthnot.
Reeves, for example, has no classification for Arbuthnot's much

emphasized concept of the translatory lie. Moreover, as Margaret
Rose has shown, recognition of parody creates an enormously
complicated hermeneutic for the reader to mediate.[96] In this case we
should certainly be careful before inferring a neat symmetry of
message between the texts – that which Reeves directly condemns,
Arbuthnot condemns similarly through ironic praise. Nevertheless,
in some other respects the similarities between the two texts are
actually more extensive than Carstens allows, and in terms of the
context of a preoccupation with lying and casuistry, much stronger.

Both works derive truth from God and lies from the Devil.[97] Each
offers an initial definition. Whereas Arbuthnot's definition is in the
casuistic idiom with its emphasis on malicious intention, Reeves'
definition is full square in the Augustinian tradition as mediated,
for example by Clarkson, and in strikingly similar phrasing. A lie
is not a mistaken notion nor a logical untruth, for a logical truth
may be a moral lie, and it is not restricted by any criterion of evil
intent; it is speaking otherwise than we speak in the heart.[98] He
certainly notes the casuistic stress on evil intention, but for Reeves,
this is not its essence for if it were, lying could be mitigated, a point
made also by Clarkson.[99] In Reeves' sermon and in *The Art of
Political Lying* it is claimed that truth is easy and straightforward
whereas lying requires art; each work conventionally expresses the
view that lying has much increased of late, 'state slander' as Reeves
put it 'is a growing evil', and Arbuthnot's Projector refers to many
recent inventions.[100] In each we find a distinction between calumny
and 'whisper'. For Reeves there is little difference between them.[101]
In *The Art of Political Lying* the distinction is mentioned to draw
attention to the different velocities at which lies travel in the public
domain.[102] Those who do most to spread lies, writes Reeves, are
removed from the centre of government and so see only 'through a
glass darkly'.[103] There is here a shared image of light and vision,
and again it had been one used in *The Examiner*. Reeves' point,
however, seems to suggest the division of labour between the
invention of lies and the organs of propagandistic distribution that
we find in the 'Pseudologia'. Reeves asks rhetorically whether
conversation nowadays is any more than 'a match at fencing . . . ?'
and the Author of the 'Pseudologia' explains politicians' belief in
their own lies in terms of mutual conversation being too combative.[104] For Reeves, in stark contrast to the casuists, there is never a
necessity to lie, whatever good we may think comes from lying, it
is always a sin and must be seen as that. He provides an explicit

attack on the (unnamed) casuists (mostly Catholic, but among them the odd Protestant);[105] in doing so he leans on the authority of Augustine's *De mendacio* which may be to take aim at the relatively recent republication of Daniel's *Discourses*.[106] Reeves goes so far as to claim that doctrines of equivocation and reservation 'spun out of the art of Leasing [sic]' obviate the need for the devil to tell lies for his own ends.[107] As Reeves abridges the casuists, ' it is no sin to deceive him by a Lye, who has no Right to Truth'.[108] By an argumentative *reductio*, Clarkson had seen such a possibility as a requirement of Catholicism *per se* in its treatment of heretics and it becomes by a further extension a basic proposition of the 'Pseudo-logia', chapter three.[109] In politics, we are suavely informed, no one has any right to truth and lying is the only virtue and necessity. It is at this point that the Projector demurs with the 'seeming difficul-ties' raised by Reeves' sermon.[110] And just as the 'Pseudologia' assures us that effective lying unifies *honestas* and *utilitas*, so Reeves assures us that honesty is always the best and most advantageous policy.[111]

In juxtaposing Reeves' sermon and the *Art of Political Lying* two other texts have become significant, Gabriel Daniel's recent and powerful defence of casuistry against what he predictably claimed to be the lies of the Jansenists and Clarkson's compendious attack on Catholicism. These two works were in fact in agreement on one central issue, that Jesuit casuistry was not atypical of Catholic doctrine. Yet, whereas Daniel argued this in order to legitimate casuistry within the Church, hence his deference to de Condren, Clarkson took it to be at the symbolic centre of Catholic moral corruption. In many respects, as I have indicated, Reeves' argument echoes and may have drawn on Clarkson and the general modus operandi typically employed by Clarkson is what had offended Daniel: the Jesuits are set up as straw men, the real arguments they present and the strict rules of limitation they insist upon in mitiga-tion are all ignored, their critics proceed by a form of *reductio*.[112] There are bounds he insisted, Jesuits are not setting up new rules of morality.[113] It is exactly this complaint, levelled against Pascal, that might have been made against Clarkson who had proceeded, however learnedly by process of *reductio*; an exception is taken to establish a new a rule, or because there are no predetermined bounds to casuistry, then by implication anything is encouraged. Perjury possible becomes perjury required, lying becomes part of morality.[114] Again, as lying is seen by Catholics as harmful by degrees, it can be construed by a sorites argument as not harmful, there being only a fractional difference between being benign, marginally harmful and so on.[115] It is a similar process we find in *The Art of Political Lying* but for ironic rather than directly polemical attack; what Reeves finds wrong and potentially dangerous, the Author and Projector posit as right and necessary, creating, in effect the new rules of which the casuists were accused.

Carstens remarks that what is also significant is the lack of parallel between *The Art of Political Lying* and the latter part of Reeves' sermon, which is an unqualified exhortation to follow the principles of honesty and sincerity, for lying destroys society. As Reeves puts it, lying is 'High Treason against the Fundamental Law of Society' and a sin against the profession of Christianity.[116] Thus Carstens concludes, 'the disapproval of the Church stands against everything [the Projector and Author] say, and Arbuthnot intends the reader to remember it.'[117] If the sermon was being used as a parodic pre-text for *The Art of Political Lying*, then the silence about sincerity may be significant and the allusive presence of Reeves' sermon as the only text within *The Art of Political Lying*, invites a counter-point between the two works. But there is a sense in which the force of the whole sermon parodically inverted is already incorporated into the expressed principles of the 'Pseudologia': Lying is the best and only policy, it is possible to have a society of liars, it is what we call a political system. I have already drawn attention to what seems to be an undercurrent of allusion to Hobbes and the state of nature (ch. 7, V) and this is altogether closer to the surface in the latter part of Reeves' sermon. It is because man and language are social and because 'fixing the signification of words and keeping to . . . current meaning' is so necessary that lies are such high treason. If the state of nature really were a state of war, he remarks, it would not matter how we carried on; but if civilization is anything it must have sincerity and truth. Government turns on the fidelity of the tongue.[118] It is precisely this which the 'Pseudologia' inverts. And the invitation to contemplate Reeves' sermon provides a surrogate set of allusions.

Even if this is so, however, a ringing endorsement of Reeves' simple dichotomies between truth and falsehood and his ritualistic reiteration that lying destroys society does not neatly follow. Intending the reader to remember the condemnation of the Church is not necessarily requiring the reader accept it. What I have hypothesized as Arbuthnot's satiric strategy should lead us to expect something more subtle. Given the conceptually extensive notion of lying as it is presented in *The Art of Political Lying*, to cover all and every statement or gesture in the political system, lying also clearly includes the lubricants as well as the solvents of social intercourse. Leaving aside the casuistic riposte that one can imagine occasions on which the survival of society might depend upon a certain mealy mouthedness with the truth and that asserting less than the truth could accord with high social values,[119] Arbuthnot was enough of a court figure to understand the necessity of politeness and social grace. The Queen's doctor was familiar enough with different Addisonian bow sizes; and like La Rochefoucault, that greatest of analysts of the uses of courtly hypocrisy, he recognized the social importance of some degree of disingenuousness.[120] Discretion and diplomacy might be impossible to disentangle from Reeves' notion of lying just as they form part of Addison's seamless web of education in French Statecraft. It is above all this which is vital in the notion of promissory lies, embracing Addison's meaningful gestures of innuendo, evasion, ingratiation; and these provide the shared codes which rather than destroy, delineate a society. It is particularly clear from the hints we are given, that the image of political society imagined by the 'Pseudologia' involves a rigid encoding of decorum and functional roles. Arbuthnot's political world of word and gesture is, in short, a world of politeness in the recognized eighteenth-century sense. It was seen as the fundamental social lubricant, La Rochefoucault's *politesse* which Abel Boyer felicitously called the 'dexterous management of our Words and Actions, whereby we make other people have [a] better Opinion of us and themselves'.[121] Lying presupposes and requires forms of socialization, and so expresses a sense of civilization. The question raised by a contemplation of *The Art of Political Lying* in conjunction with Reeves' sermon is whether total honesty and sincerity is as corrosive of political life as the 'Pseudologia's total dishonesty.

IX

An altogether more significant omission from the sermon than the peroration on total honesty concerns the clear prohibition on lies told for delight. Such lies form a distinct classification for Reeves. They are not as pernicious as malicious lies, especially the state lies of politics, but they remain lies and so evil in themselves.[122] Reeves excludes only from his censorious beam of sincerity, New Testament parables, for these are not to be believed or taken literally. It is a moot point how he would have regarded *John Bull*. Or rather it probably isn't, for as satire *John Bull* was in large measure casuistic. Satire might delight but it might well be cruel and could defame for the greater good of censure and reform. The satirist's love of truth as an end did not prohibit all sorts of means for getting to it. Reeves will have nothing to do with such Jesuitical (or Quintilianesque) justifications of evil done for good. As he dolefully laments, much truth has been sacrificed to the idol of delight.[123] The sermon's peroration then, exhorts an unqualified High Church Augustinian adherence to the principle of truth at all times, which no case can modify, no casuistry compromise. Against this stands the 'Pseudologia'. As I have suggested in other respects, Arbuthnot's relationship to the 'Pseudologia' is not transparent and we may suspect that he does not side unequivocally with William Reeves any more than the Church is made to stand *equally* against Projector and Author. The requirement to adhere, regardless of circumstances, to sincerity and truth, to the unity of word and heart must have made Arbuthnot feel uncomfortable if he had been swayed by Reeves' unrelenting Bayleful austerity. But perhaps the difficulties noted by the Projector were 'seeming'. Daniel had already provided a case for saying as much, pointing out that Abraham, Christ and Augustine had all on occasion made use of reservation or equivocation for some greater good.[124] For if the 'Pseudologia' and its Projector provide a backhanded compliment to William Reeves through a reinforcing parody of his attitude to state lies, and the criticism of The Queen in particular, *The Art of Political Lying* is itself an example of a lie partially for delight, indeed for the casuistic good end of satiric censure and reform which could lay so close to preaching.

True to the rhetorical principles to which Arbuthnot alludes, there is a serious doctrine to be communicated through delight. This I believe has two distinguishable aspects; it involves the invitation to question just how far the political is constituted by rules of lying, which I shall discuss in the following chapter; and it intimates the desire to arm the reader, the political participant against those following the regulatory rules of lying. In short, one may hypothesize that the dual motivation governing *The Laws of Chance*, to understand the rules of probability and deal with cheats, is carried through to *The Art of Political Lying*. It is through the classification of plausible lies that the reader is alerted and given a critical way of organizing responses to political discourse.[125] As a hand book for the promising politician, it is a Greek gift. Here again we have something of a variation upon Machiavelli's *Prince*, at least in the most likely form in which Arbuthnot knew it. He owned a copy of Neville's translation which includes Neville's own understanding of *The Prince* presented in the form of a fake letter by Machiavelli in his own defence, the lie gaining a garnish of plausibility from the printer John Starkey who provided a spurions provenance.[126] It is exactly Neville's point that so far from teaching evil, Machiavelli is teaching us, and so helping us morally to cope with how bad princes in fact behave, it censures and warns. The work is in this sense a 'satyre'.[127] Arbuthnot's reductive pattern of argument, would thus lead to the conclusion that Neville's bad prince who rules by fraud and plays the fox is the norm among all politicians, and the 'Pseudologia' by adumbrating and abridging the rules of dishonesty is educating us all and is in Neville's sense a 'satyre'.

As we have seen, one part of the genesis of *The Art of Political Lying* may have been in the court jest Arbuthnot and Swift played on the Queen's ladies in waiting, a lie which Reeves might well have classified as malicious (see above, ch. 3, I). Generally Arbuthnot had a fascination with rumour and he and Swift started one or two simply to see what would happen. In April 1713, they would be found 'contriving a Lye for the morrow' simply to experiment, as Arbuthnot might have re-described it, with the 'celerity' of motion.[128]

The liberty to preach, Arbuthnot later reflected to Swift, must make any man happy, 'if it did not border on simony, I could really purchase it for a sum of money'. How he would relish the opportunity of 'venting' himself 'to a congregation once a week'.[129] The double entendre of 'venting' (as in the Author's scheme that the Whigs might 'vent' truth) indicates a certain critical irony towards the duties of Swift's vocation so fulsomely fulfilled by William Reeves; and in a way that would have doubly offended the good sermoniser, Arbuthnot had already taken liberties with preaching consistent with the doctrines of *The Art of Political Lying*.

X

Arbuthnot's one serious foray into political controversy had been a bogus sermon involving salutary falsehood for the good end of Union between England and Scotland. In fact, the *Sermon Preach'd at Mercat Cross* contains more than one form of lie within the terms discussed in the 'Pseudologia'.[130] To begin with, it was never preached at Mercat Cross or anywhere else. The sermonic form is assumed to enhance the authority of the message to Scots, ever suspicious of England and fearful of the clear loss of independence Union would bring. By adopting this form, Arbuthnot insinuates the principal falsehood involved that the speaker is a Scottish Presbyterian minister.[131] It is what the Author of the 'Pseudologia' would call an 'additory lie' and as Arbuthnot was an Episcopalian Scot, it was within the bounds of probability. This not only translates the authority of the Kirk but enables him to maintain the intimacy of appeal appropriate to a sermonic venting and so to structure the argument around various anaphoric formulations of 'us' and 'them'. By turns, 'we' are Scots, believers in Union and co-religionists; 'they' are English or they are opposers of Union, whose arguments are 'frivolous' or 'founded upon gross false-hoods'.[132] This allows the ambiguous reference to 'our trifling differences in religion'.[133] It is, at least, grossly disingenuous, as was the claim that 'the religion of the Church of England is imposed upon no man within the dominions of England'.[134] Formally this was true but there were considerable disadvantages to not being of the Church of England, and Arbuthnot was not one to have them eased. As, throughout the sermon, Arbuthnot is stressing the practical advantages of Union, this amounts to what Reeves would call 'a moral lie' and 'a lie before God'.[135]

Mention has been made of *A Sermon at Mercat Cross*, not to conclude that a dishonest John changed his ways, or even that with respect to political honesty, Arbuthnot failed to practice what he preached, or rather didn't preach. The situation is altogether more complex than a Reeves-like mechanical rectitude will allow, or which a truth-table vision of political discourse can accommodate. If in other respects *The Art of Political Lying* takes up and plays with views Arbuthnot took seriously, there is no reason to exclude *Mercat Cross*. The work displays a sustained and structurally significant degree of disingenuousness for a cause in which Arbuth-not believed and which he presented with sufficient power for it to

be re-printed to re-enforce the benefits of Union in 1745.[136] There
is in a word, something casuistic about his sermon, but subtly so.
He is asking the Scots to depart from principle and recognize
self-interest and a higher public good. The falsity asserted in the
title is thereafter suggested and the reader encouraged to draw the
erroneous conclusion about 'our' religious differences; or rather, in
context the equivocal pronoun allows re-specification according to
who one is, which side of the border one stands. In this way there
is a structural similarity with the suggestive open-endedness of *The
Art of Political Lying*, evoking, to recall de Condren, the similarly
amphibolous 'your' Dutch or English lie, and inviting creative
application which could hardly stop at questioning the whole
nature of public discourse when seen in terms of truth and lies. To
return to the whole issue of moral principle and specific case: in
juxtaposing the reductively casuistic doctrines of the 'Pseudologia'
with the unbending anti-casuistic principles of Dr Reeves, Arbuthnot, true to his satiric style, may have been inviting the reader to
negotiate a route between Scylla and Charybdis. His aim, then
would have been to provoke a greater critical self-awareness of
what it is to say anything in a world in which men neither behave
or speak as they should;[137] and in part to do this on the basis of
his own foray into the public sphere, a piece of deliberative
rhetoric which was only 'in the main honest' and 'plain dealing'.[138]
To explicate further the possibility that the lie may in fact function
to achieve a surer grasp of truth, we have to turn briefly to the
notion of a philosophic lie, located somewhere between malice and
delight.

NOTES

1   Anthony Copley, *An Answer to a Letter of a Jesuitical Gentleman*
    (1601) p. 199. See Perez Zagorin, *Ways of Lying, Dissimulation,
    Persecution and Conformity in Early Modern Europe*, (Camb. Mass.:
    Harvard University Press, 1990), pp. 196–9 for Copley among other
    Catholic objectors to casuistry.
2   Thomas Morton, *A Full Satisfaction Concerning a Double Romish
    Iniquitie* (1606), p. A4; Isaac Causabon, cited in Zagorin, *Ways of
    Lying*, pp. 203–4; Henry Mason, *The New Art of Lying Covered by
    Jesuits under a Vaile of Equivocation*, (1624).

3   Robert Burton, *Anatomy of Melancholy*, Intro. Holbrook Jackson, (London: Dent, 1961 edn.), vol. 1. p. 65ff; the belief that it was an age of dishonesty is found in Stephen Zwicker, *Politics and Language in Dryden's Poetry: The Arts of Disguise*, (Princeton: University Press, 1984), Introduction; Perez Zagorin, *Ways of Lying*.

4   Zagorin, *Ibid.*, p. 10ff.

5   Mason, *The New Art*, pp. B2, 25, 29, 37.

6   Johann Sommerville, 'The New Art of Lying', in Edmund Leites, ed. *Conscience and Casuistry in Early Modern Europe*, (Cambridge: University Press, 1988), pp. 160–1.

7   Cicero, *De oratore*, ed. and trans. E. W. Sutton and H. Rackham, (Camb. Mass.: Harvard University Press, 1942) I.56.239; I.50–2, on eloquence rather than true knowledge winning arguments; *De particione oratoria*, xxviii on the rules and tactics of arguing on both sides. See also III. 21.

8   Quintilian, *Institutio oratoria* ed. and trans. H. E. Butler, (Camb. Mass.: Harvard University Press, 1920), II, 17, 19, 21.

9   Sommerville, 'The New Art of Lying', p. 161; Toulmin and Jonsen, *The Abuse of Casuistry*, (Los Angeles: University of California Press, 1988), pp. 196–7; C. Jan Swearingen, *Rhetoric and Irony: Western Literacy and Western Lies*, (Oxford: University Press, 1991), ch. 5.

10   Jan Swearingen, *Ibid.*

11   See for example, Zagorin, *Ways of Lying*, p. 18.

12   Aquinas, *Summa theologiae*, 2.2. ques. 110, art. 4. See Zagorin *Ibid.*, pp. 29–31.

13   Zagorin, *Ibid.*, pp. 12ff, 34ff.

14   See Sissela Bok, *Lying: Moral Choice in Public and Private Life*, (Sussex: Harvester, 1978), ch. 1; John Barnes, *A Pack of Lies: Towards a Sociology of Lying*, (Cambridge: University Press, 1994). pp. 6–10.

15   Arbuthnot, *APL*, p. 8; cf Richard Steele, *The Spectator*, 352, 14, April, 1712.

16   This is Augustine's principal understanding reiterated for example by David Clarkson, *Practical Divinity of The Papists* (1675), ch. 8. sect. 9, pp. 245ff; for further examples see Sommerville, 'The New Art of Lying', p. 161.

17   Sommerville, *Ibid.*, p. 161; Barnes, *A Pack of Lies*, pp. 10–19 for a succinct discussion.

18   Bok, *Lying*, p. 14ff.

19   Cf Morton, *A Full Satisfaction*, A3 and Clarkson, *The Practical Divinity*; with Daniel, *Discourses of Cleander and Eudoxus* (1694, 1704) pp. 443–5.

20   See Barnes, *A Pack of Lies*; and Bok, *Lying* at length.

21   Sommerville, 'The New Art of Lying', at length; Toulmin and Jonsen, *The Abuse of Casuistry*, pp. 250–202.

22   Morton, *A Full Satisfaction*, A4, pt. 3 ch. 15; Mason, *The New Art*, ch. 1, p. 21 ch. 2, pp. 32–3.

23   Morton, *A Full Satisfaction*, A4.

24   Gaines Post, *Studies in Medieval Legal Thought: Public Law and the State*, (Princeton: University Press, 1964), esp. 260ff; but see also Alan

Gilbert, *Machiavelli's Prince and its Forerunners*, (Durham: Duke University Press, 1939) for a chapter by chapter listing of partial antecedents.

25 Machiavelli, *The Prince*, ed. and trans. Russell Price and Quentin Skinner, (Cambridge: University Press, 1988), chs. 15–19.

26 Thomas Fitzherbert, *An sic utilitas in scelere*, (Rome 1610).

27 Justus Lipsius, *Politicorum sive civilis doctrinae libri sex*, (1589); see Zagorin, *Ways of Lying*, p. 124.

28 Thomas Hobbes (?) 'A Discourse Upon the Beginning of Tacitus' *Horae subsecivae*, (1620) *Three Discourses*, ed. Noel B. Reynolds and Arlene Saxonhouse, (Chicago: University Press, 1995), p. 53.

29 Michel de Montaigne, 'Of Liars', *The Complete Essays*, trans. Donald M. Frame, (California: Stanford University Press, 1965 edn.), p. 24; Sir Thomas Browne, *Christian Morals* printed with *Religio medici*, ed. Henry Gardiner, (London, 1845), 2.3, p. 287.

30 Erasmus, *Paraphrases in Novum Testamentum; Annotationes*, to Galatians 2: 11–14, see Zagorin, *Ways of Lying*, pp. 34–5.

31 Daniel, *Discourses*, p. 437

32 de Condren, *Traité des équivoques*, (Paris, 1643), *Oeurves complètes*, ed. Abbé Pin, (Paris, 1847–8) p. 413.

33 Mason, *The New Art*, pp. 2, 12.

34 Mason, *Ibid.*, epistle, B2.; Copley, *An Answer*, pp. 92–3; see also Robert Tynely, *Two Learned Sermons* (1609) and John King, *A Sermon Preached at Whitehall* (Oxford, 1608) each at length for for similar stress on redescriptive ingenuity. I am grateful to Dr Lori-Ann Ferrell for drawing my attention to these works.

35 Clarkson, *The Practical Divinity*, ch. 8, sect. 9, p. 251.

36 Cf Mason, *The New Art*, pp. 71–6 with Daniel, *Discourses*, pp. 413–4 on whether Abraham lied to the Egyptians over his relationship with Sarah.

37 See John Wilson, *Politically Speaking The Pragmatic Analysis of Political Language*, (Oxford: Blackwell, 1990), esp. ch. 6.

38 Barnes, *A Pack of Lies*, p. 140, quoting Blake.

39 Navarrus, (Martin Azpilcueta), *Enchiridion*, (1549) which became something of a teaching manual for Jesuits. See Zagorin, *Ways of Lying* pp. 165–70; *APL*, p. 8 for allusion to mixed modes of Lying.

40 Henry Garret, *A Treatise of Equivocation* (1595); Robert Persons, *A Treatise Tending Towards Mitigation*, (1607).

41 Morton, *A Full Satisfaction*, pt. 3, pp. 47, 85; Mason, *The New Art*, ch. 1; see Sommerville, 'The New Art of Lying' p. 179.

42 John Barnes, *Dissertatio contra aequivocationes*, (Paris, 1625), see Sommerville, 'The New Art of Lying', p. 180; cf. John Barnes, *A Pack of Lies*.

43 Mason, *The New Art*, is fastidious in excluding Catholics generally from his sights; Clarkson *The Practical Divinity*, is totally inclusive.

44 Clarkson, *Ibid.*, ch. 5, p. 121. For a fine discussion of the Jansenist attack on the Jesuits' distinctions in the concept of sin see John Kilcullen, *Sincerity and Truth.*, (Oxford: Clarendon Press, 1988) Essay 1, pp. 7ff.

45 For a account of the problem of *paradiastole*, see Quentin Skinner, *Reason and Rhetoric in The Philosophy of Hobbes*, (Cambridge: University Press, 1996) pp. 142ff; 161ff; 174; also, Barnes, *A Pack of Lies* pp. 115ff.
46 Plato, *Republic*, 560 E- 561. Paul Shorey aptly translates as 'euphemistically denominates'. See *Republic*, trans. P. Shorey, (Camb. Mass.: Harvard, 1970 edn.) vol. 2, p. 299.
47 Mason, *The New Art*, B2, p. 12.
48 L. Anderton, *Remarks on the Present Confederacy*, (1693) in *Somers Tracts*, vol. 3, first collection, (1748), pp. 545.
49 Montaigne, *Essays*, 1, p. 150ff.
50 John Earle, 'The Character of a Sceptic', *Microcosmographia* (1626); see also Brown, *Religio medici*, sect.1. See also Samuel Butler *Characters*, ed. C. W. Daves, (Cleveland: Case Western Reserve Press, 1970),'A sceptic', p. 165.
51 The legitimate exploitation of these is discussed by Cicero, *De partitione*, xxx, 107–8.
52 Mason, *The New Art*, ch. 1, p. 3ff, 31ff; Aristotle, *On Sophistical Refutations*, 165b, 168a, 169a vii has discussion of equivocation all of which deal with it as a linguistic and therefore accessible problem, so setting a very solid precedent for Mason's distinction between the old and inevitable and the new and noxious types of equivocation.
53 Hobbes, *Leviathan*, chs. 13, 18.
54 Skinner, *Reason and Rhetoric*, pp. 317ff.See also Locke, *Essay*, bk. 2 ch. 28, 10–12 for a similarly Hobbesian point.
55 Jonathan Swift, *Gulliver's Travels*, 3, ch. 5.
56 Cicero, *De partitione*, ii, 5.
57 As Locke points out in the *Essay*, a system of understandings works if ideas are conformable to expectations, they do not have to be true, bk. 2, ch. 34, 4.
58 Arbuthnot, *APL*, pp. 13, 12; Steele, *The Spectator*, 352, 14, April, 1712.
59 Arbuthnot, *Ibid.*, p. 14.
60 Arbuthnot, *Ibid.*, pp. 14–15. The raw head is possibly an allusion to the head allegedly found under the Capitol in ancient Rome and signifying the city's foundation in blood.
61 '*La pittura e una poesia muta e la poesia e una pittura cieca*', Leonardo, quoted in R. Hinks, *Myth and Allegory in Ancient Art*, (London: Warburg Institute, 1939), p. 14, from Horace, *Ars poetica*; Ben Jonson, *Timber* (1641) 'Poesis et pictura', 'It was excellently said of Plutarch, poetry was a speaking picture, and picture a mute poesy'.
62 Quintilian, *Inst. orat.*, II, 17, 21; also II, 14, 5.
63 Quintilian, *Ibid.*, II, 18, 1. It is noteworthy also, that the 'illustrious Person' in the Academy of Legardo responsible for so many delusory schemes is called the 'universal Artist', Swift, *Gulliver's Travels*, 3, ch. 5.
64 William Dunbar, 'Lament for the Makaris' in The *Poems of William Dunbar*, ed. W. Mackay Mackenzie, (London: Faber, 1932), pp. 20–3; see also Tynely, *Two Learned Sermons* for the continued association of linguistic re-description with physical transformation.

65 I am much indebted to Craig Muldrew on this point; also, J. G. A Pocock, *The Machiavellian Moment*, (Princeton: University Press, 1975), p. 487.

66 Arbuthnot, *APL*, pp. 12–13.

67 Locke, *Essay*, bk. 2, ch. 27, 10, 11–14; Christopher Fox, *Locke and the Scriblerians: Identity and Consciousness in Early Eighteenth-Century Britain*, (Los Angeles: University of California Press, 1988), pp. 2–18; Toulmin and Jonsen, *The Abuse of Casuistry*, p. 222; Condren, 'Liberty of Office', forthcoming.

68 Peter M. Briggs, 'Locke's *Essay* and the Strategies of Eighteenth-Century English Satire,' *Studies in Eighteenth-Century Culture*, 10 (1981), p. 135ff, Arbuthnot is not discussed.

69 For an illuminating discussion of the double identity and marriage/bigamy episode in the life of Martinus Scriblerus see Fox, *Locke and the Scriblerians*, p. 96. The immediate inspiration for the *topos* may have been the exhibition in London of a remarkable pair of siamese twins in 1708, or more generally an often gullible fascination with freak shows. See Pat Rogers, *Literature and Popular Culture in Eighteenth-Century England*, (Sussex: Harvester, 1985), pp. 12–13. On the association of personality and money see, Lawence Klein, 'Property and Politeness in the Eighteenth-Century Whig moralists', in John Brewer and Susan Staves, eds., *Early Modern Conceptions of Property*, (London: Routledge, 1995), pp. 222 ff; Pocock, *The Machiavellian Moment*, ch. 13.

70 Locke, *Essay*, for example, bk. 3, 9, 6.

71 Barnes, *A Pack of Lies*, p. 69.

72 Locke, *Essay*, bk. 3, 9, 6.

73 Pierre Bayle, 'John Fox', *Commentaire philosophique sur ces paroles de Jesus-Christ, 'Contrain-les d'entrer*, (1686, 1688), on which see Kilcullen, *Sincerity and Truth*, 2, esp. pp. 64–6.

74 *The Cheating Age*, (1712), is robust in its claims.

75 For a discussion see Conal Condren, 'Andrew Marvell's *Account of the Growth of Popery and Arbitrary Government*, in C. Condren and A. D. Cousins eds., *The Political Identity of Andrew Marvell* (Basingstoke: The Scolar Press, 1990).

76 Charles Davenant, *The True Picture of a Modern Whig*, (1701), pp. 4, 6.

77 John Dunton, *Athenian Sport*, (1707), paradoxes 79, 73, 28.

78 Swift, *The Examiner*, 14, 9 Nov. (1711), p. 9.

79 *The Plain Dealer*, 5, 10 May, 8, 31, May 9, 7 June, (1712).

80 *The Plain Dealer*, 14, 12 July (1712).

81 Joseph Addison, *The Spectator* 305, 19, Feb, (1712); see Lester M. Beattie, *John Arbuthnot Mathematician and Satirist*, (Camb. Mass.: Harvard University Press, 1935, and New York: Russell and Russell, 1967), p. 290.

82 Daniel, *Discourses*, p. 444, 415.

83 Addison, *The Spectator*, no. 305; cf. *The Post Boy*, 2192, 11–13 Jan. (1708/9), for a genuine advertisement; Beattie, *John Arbuthnot*, pp. 296–7, on Addison and Arbuthnot.

84 Davenant, *The True Picture*, pp. 5–7.

85   Patricia Carstens, 'Political Satire in the work of John Arbuthnot,' London University Unpublished PhD thesis (1958), p. 298.

86   Arbuthnot, *John Bull, John Bull in His Senses*, ch. 1, p. 26.

87   Arbuthnot, *Ibid.*, p. 25. This may be an allusion to the papers found in Algernon Sidney's cabinet, which were used against him in his trial, they were later published as *Discourses on Government*.

88   Arbuthnot, *An Appendix to John Bull Still in His Senses*, ch. 3, p. 87.

89   Arbuthnot, *John Bull, Law is a Bottomless Pit*, ch. 5, p. 9.

90   Arbuthnot, *John Bull, Lewis Baboon Turned Honest*, ch. 4, p. 111.

91   Arbuthnot, *APL*, p. 9.

92   William Reeves, *The Nature of Truth and Falsehood*, (1712); see Carstens, 'Political Satire', p. 322.

93   Carstens, *Ibid.*, p. 324.

94   Carstens, *Ibid.*, p. 322.

95   Reeves, *Truth and Falsity*, pp. 5–7.

96   Margaret Rose, *Parody / Meta-Fiction*, (London: Croom Helm, 1979); and in more detail, *Parody, Ancient, Modern and Post-Modern* (Cambridge: University Press, 1993).

97   Reeves, *Truth and Falsity*, pp. 10–11, 18.

98   Reeves, *Truth and Falsity*, pp. 5–6; cf Clarkson, *The Practical Divinity*, ch. 8, sect. 9, p. 245ff.

99   Clarkson, *Ibid.*, p. 245.

100 Reeves, *Truth and Falsity*, p. 8; see also Swift *Examiner*, 14, 9 Nov. (1710); Steele, *The Spectator*, 352, 14, April, (1712)

101 Reeves, *Ibid.*, p. 6.

102 Arbuthnot, *APL*, p. 20.

103 Reeves, *Truth and Falsity*, pp. 7–8.

104 Reeves, *Ibid.*, p. 18; Arbuthnot, *APL*, p. 19.

105 Reeves, *Ibid.*, pp. 10, 16ff.

106 Reeves, *Ibid.*, pp. 13, 16–17.

107 Reeves, *Ibid.*, p. 7; cf. John Barnes, *Dissertatio*, and the discussion in Sommerville, 'The New Art of Lying'.

108 Reeves, *Ibid.*, p. 10; and. Morton, *A Full Satisfaction*, pt. 3, p. 89; Daniel, *Discourses*, 7, p. 437 asserts just this point.

109 Clarkson, *The Practical Divinity*, ch. 7, sect. 4, p. 204.

110 Arbuthnot, *APL*, p. 9.

111 Reeves, *Truth and Falsity*, p. 24; Steele, *The Spectator*, 352, 14, April, (1712).

112 Daniel, *Discourses*, 3, at length.

113 Daniel, *Discourses*, pp. 101, 108.

114 Clarkson, *The Practical Divinity*, pp. 204, 245.

115 Clarkson, *Ibid.*, p. 248.

116 Reeves, *Truth and Falsity*, pp. 15, 18; cf. Daniel, *Discourses*, p. 412.

117 Carstens, 'Political Satire', p. 325.

118 Reeves, *Truth and Falsity*, pp. 15, 16.

119 Daniel, *Discourses*, 7, p. 425ff.

120 See Beattie, *John Arbuthnot* on Arbuthnot's negotiations with Flamstead; George Aitken, *Life and Works* of John Arbuthnot, (Oxford: Clarendon Press, 1892), pp. 36–7.

121 Abel Boyer, *The English Theophrastus*, (1702), p. 106; cited E. Hundert, *The Enlightenment's Fable Bernard de Mandeville and the Discovery of Society*, (Cambridge: University Press, 1994), p. 119.
122 Reeves, *Truth and Falsity*, p. 9; cf. Clarkson, ch. 8, sect.9, p. 246.
123 Reeves, *Ibid.*, p. 8.
124 Daniel, *Discourses*, p. 415ff.
125 John Barnes, *A Pack of Lies*, p. 108ff.
126 Henry Neville, *The Works of the Famous Nicholas Machiavel*, (1675), unpaginated.
127 Neville, *Ibid.*
128 Arbuthnot, *APL*, p. 20; see Alexander Ross, 'The Correspondence of John Arbuthnot', Cambridge University, unpublished PhD thesis, (1956), vol. 1, p. 44.
129 Alexander Ross, 'The Correspondence of John Arbuthnot', vol. 1, p. 361.
130 *A Sermon Preach'd to the People at the Mercat Cross of Edinburgh, on the Subject of the Union* (1706), in Aitken, *Life and Works*, p. 396ff.
131 Arbuthnot, *Ibid.*, p. 396. The lie was sustained behind the printed page, see Ross, 'Correspondence', vol. 2, letter 10, (1706–7), p. 911.
132 Arbuthnot, *Ibid.*, p. 406.
133 Arbuthnot, *Ibid.*, p. 407.
134 Arbuthnot, *Ibid.*, p. 407.
135 Reeves, *Truth and Falsity*, p. 5.
136 Aitken, *Life and Works*, p. 392.
137 Machiavelli, *The Prince*, ch. 15.
138 Arbuthnot, *John Bull, Law is a Bottomless Pit*, ch. 5, p. 9.

8 Philosophical Lying

'*Philo*, with twelve yeares study, hath been griev'd
To be understood; when will hee be beleev'd?

John Donne, epigraph, *Satyres* 1633.

'The Antithesis or See-Saw,
whereby Contraries and Oppositions are balanc'd in such a way, as to
cause a Reader to remain suspended between them, to his exceeding
Delight and Recreation'.

John Arbuthnot and Alexander Pope, *Peri Bathous*, p. 50

I

I have argued that John Arbuthnot, even if not acting the preacher
and venting himself upon the public, did have a serious doctrine to
communicate in *The Art of Political Lying*. It was not simply that
we should be honest and sincere in all things. Part of it was a direct
extension of Neville's reading of Machiavelli's *Prince* as satire. If
the 'Pseudologia' is there to instruct the politicians, the account we
are given of it arms us against their activities. Yet, as public opinion
is itself part of the political system for Arbuthnot, both teaching
and fore-arming apply to us all. This is to introduce more than a
tincture of paradox and through paradox it raises the issue of just
how far forms of dishonesty provide the constitutive rules of
politics. At what point between an insight and a reduction to an
extreme do we have something like the truth? It is Niklas Luhmann
who in recent years has become celebrated for claiming a necessary
interdependence of forms of self-reference and the notion of a
system.[1] Arbuthnot was not attempting anything of comparable
scope, but the notion of a political system that arises in his work,
like Luhmann's is derived from metaphors of circulation and
exchange. Moreover, it is also completed by a precise form of
self-reference. So too, what I have hypothesized as Arbuthnot's
potentially inclusive sense of satire is supported by the paradoxical
structure of his work.

Paradox is a term which from the late sixteenth century has been
given to unhelpfully broad and slippery use.[2] There are several
senses relevant to the text which need disengaging. First, there is

144

the probability paradox where the word may refer simply to a statement or argument which is or is believed to be against common opinions, *para-doxai*. It is in this sense that Socrates uses it in *The Republic* in advocating that philosophers become kings.[3] It is in this sense also that Cicero elaborates a series of paradoxes, in his *Paradoxa stoicorum*, such as that only the wise are truly rich.[4] This work became something of a model for Renaissance discussions. Yet how far something is paradoxical in this Platonic or Ciceronian way is not self-evident. Something probable is being claimed, often involving some sort of moral aspersion on common opinion. It may by degrees be true or false. In this first sense, it is not clear how far Arbuthnot's *Art of Political Lying* is a paradox, given the predisposition to believe that politics is a domain of falsehood, the art, as Isaac Disraeli was to put it, of governing mankind by deceiving them.[5] This, as the eighteenth century can show us, was no latter day cynicism and even Plato who despised the sophists and the tricks they played with the truth founded his ideal *polis* on noble lies.[6] Nevertheless, in so far as Arbuthnot presents an argument that politicians should not be honest, the work is paradoxical in this first sense. We do normally hope that they will be honest, but more to the point, those engaged in politics continue to encourage expectations of honesty, if only through the ritualistic shock they exhibit, now as in the eighteenth century, in the face of mendacity. A qualification then, remains in order. If *The Art of Political Lying* is parodoxical in advocating dishonesty, it was quite in tune with the dogmas of common opinion in expressing the view that like it or not politicians almost invariably lie. The full consequences of this tension will become apparent in discussing a third sense of paradox.

In the meantime, a second sense of paradox, being an extension of the first, needs noting. This is a paradox as an unlikely defence of the seemingly indefensible or odious, one of the central topics of epideictic rhetoric. If the defence of the indefensible is successful, the sense of paradox becomes acute, but merely to attempt it is to run against common opinion. It is in this second sense that Erasmus's *In Praise of Folly* is a highly satiric paradox. Similarly, Sir William Cornwallis wrote a series of *Essays or Paradoxes*, which are forms of epideictic rhetoric with a varying sharpness of satiric edge.[7] As I have claimed, Arbuthnot's *Art of Political Lying* does have a number of epideictic features (ch. 5, II) and it is clear that the combined voices of the Author and Projector are producing something paradoxical in this second sense.[8]

II

It is a third sense of paradox, however, generally called a semantic paradox at once altogether more precise and interesting in the present context, on which I want to focus. Arbuthnot's *Art of Political Lying* is an elaborate exemplification of its most famous and difficult example. Semantic paradoxes are propositions which as Sainsbury puts it most broadly, leave us with inevitably unacceptable conclusions apparently derived from acceptable procedures.[9] In asserting a truth such a paradox involves an empirical absurdity, or a sense of indeterminacy, or self-refutation. A large part of the modern philosophical literature on-paradoxes is devoted to finding or disputing common logical features which help explain and solve them, or explaining and reformulating them so that they further resist solution. Thus, for example, for many years all semantic paradoxes were held to be distinct from logical paradoxes, such as Russell's class paradox; but more recent work has eroded the distinction and has done so in order to shed fresh light on the relationships between logic and mathematics.[10]

Many of the semantic paradoxes come down from antiquity. Zeno's paradox of Achilles and the tortoise maintains that it is impossible for Achilles to catch up the tortoise in any race between the two, allowing the tortoise a reasonable start. This paradox shares features with with probability paradoxes for it is contrary to common opinion. Indeed, we know it empirically to be false though the reasoning is superficially plausible. It is what Quine has called a falsidical paradox, one that 'packs a surprise', until we uncover its logical fallacy.[11] The most common sub-types of semantic paradox are thus likely to be classified according to the sort of logical feature they exploit. Sorites paradoxes play on indeterminacy or the slippery slope character of some classifications, such as size or weight. Diogenes Laertius records the argument, for example, that there can be so such thing as a heap of sand, since a heap is only a number of individual grains, adding one grain to another does not make a heap.[12] Meno's is a paradox of antecedents, or bi-conditionals to which I will return in the conclusion.[13] The Cretan paradox or paradox of the Liar is the most famous and logically important of this whole family and is the very model of the more general belief that lies could serve constructive purposes and even help establish a truth. Philostratus argued as much in discussing Aesop and his relationship to the poets.[14] In its early

formulations The Liar involves a contradiction arising from reflexivity, one of the most complicating features of logical structure, as
it is apt to collapse sense and reference. It is a classic example of
what Quine classifies as an antinomy, the surprise 'that can be
accommodated by nothing less than a repudiation of part of our
conceptual heritage'.[15] In its simplest form, The Liar runs thus: all
Cretans are liars, said Epimenides the Cretan. A version of this
finds its way into St. Paul, Epistle to Titus, 1.12–13. All modula-
tions of this paradox are statements which say of themselves that
they are false.[16] Modified or extended versions rely on circularity
more often than strict reflexivity. Among Jean Buridan's *Sophis-
mata* is the following: Socrates who is in Troy states that everything
Plato is saying is false; Plato who is in Athens states that everything
that Socrates is saying is true.[17] Another, which involves reflexivity
runs as follows: proposition (1) 3 + 3 = 6 (2), 6 + 6 = 12 (3),
12 + 12 = 23, (4), the number of true statements equals the number
of false ones. The crux here being whether (4) is included in the
sequence.[18] If it is, reflexivity elides sense and reference conspiring
to give us a version of The Liar. If it is not, there is no reflexivity,
no paradox, merely an arithmetical error, and we get a hint of a
notion that some semantic paradoxes might only be solved by
establishing hierarchies of logical statements, the development of a
meta-language as a necessary condition for establishing truths
about the world, a view most famously associated with the logician
Tarski.

It is this family of paradoxes, explored in Aristotle's *Sophistical
Refutations* and called *insolubilia* in the Middle Ages that has
played an important part in the development of philosophy and
logic.[19] Although they may be amusing and ingenious, as Sainsbury
remarks, their philosophical function is serious. In antiquity and in
its revival in late Renaissance Europe, Pyrrhonist scepticism was
fundamentally paradoxical in its willingness to confront any proposition with its contrary in order to stultify judgement. Ironically,
it joined with rhetoric at least in its faith in the proposition *in
utramque partem*. And we may see the Liar paradoxes as the
quintessence of the argument *in utramque partem*, offering us, if not
necessarily a Pyrrhonist proof of scepticism, certainly the rhetorician's open palm of argument in miniature. The challenge is always
to think one's way through it, as Tarski might by arguing for the
necessity of logical hierarchies. Concomitantly, whether the paradox in any or all given forms can be solved, is altogether less

important than the intellectual journey embarked upon in trying to solve it. This is exactly why The Liar has had so many formulations and has helped generate a whole modern theory of logic, dialetheism, the successor to the two truths doctrine of medieval Averroism. As the Averroistic theory of two truths held that there might be no reconciliation between theological truths and truths derived from an Aristotelian understanding of the natural world, so dialetheism holds in principle that some contradictions are logically legitimate, without our resorting to explanations, for example, through ellipsis or ambiguity and the appeal to meta-languages.[20]

Just then, as the term paradox operates as a pretty porous classifier, so too the functions of paradox are diverse.[21] Paradox could be used in the higher reaches of theology to deal with the difficulties of predicating God. By extension, I have argued elsewhere, it might be employed to specify the necessary contradictions, or tensions of complex political identity.[22] Something of the philosophical seriousness attached in particular to the Liar and its semantic cousins has rubbed off on the more miscellaneous probability and epideictic kinds of paradox which from Erasmus' *In Praise of Folly* became fashionable forms of satiric writing. Rhetoric and philosophy cannot be neatly quarantined. John Donne, introducing his paradoxes, called them 'swaggerers', telling the reader to stand up to them.[23] And I think that there is little doubt that Cornwallis' paradoxical praise of Richard III and of 'The Pocks' were serious attempts by an ex-courtier to make people think about court life and the legitimacy of contemporary politics, without explicitly seeking to direct moral responses.[24] In one, he shows fairly overtly how it is possible to redescribe a demonized villain as a noble victim; in the other, more slyly, he commends syphilis as an exclusive disease of the elite. It is something that can be hidden and rustled in silk, and so it is insinuated as a hidden presence, an invisible sign of aristocratic corruption. The English translation of Charles Estienne's *Paradoxes ce sont propos contre la commune opinion* (1593) carries front matter claiming alternately that paradoxes are simply for diversion and for serious instruction, which itself presents the reader with something of a paradox.[25] In short, with satire as an idiom of rhetoric, as well as with philosophy the point is not necessarily to accept but to think back and question.

Conversely, something of the enticing jocundity of the sophistic trick has allowed philosophers since antiquity to explore the serious through the amusing, *ridendo dicere severum*.[26] The use of wit to entice rather than excoriate has been a significant but under-explored feature of satire.[27] Here Lucian's work may well have proved a stimulus for Arbuthnot. Dryden had certainly understood it appropriately, commending Lucian's mastery of irony and his sense of decorum and arguing that his significance lay in his capacity to blend philosophy and amusement by co-opting the intellectually serious dialogue for partly comic and satiric purposes, generating laughter 'or some nobler sort of delight'.[28] In 'The Double Indictment' the claim is made that the philosophical dia-logue has been rendered more effective by being forced to smile and being paired with comedy.[29] In one dialogue Lucian explores the possibility of a society of liars, for, he claims, whole *pōleis* do tell lies unanimously and officially;[30] and in the *Verae historiae*, he effectively develops a discursive form of The Liar paradox, by calling his history true and by claiming that it is false. More honest a liar than most poets and philosophers, he writes 'though I tell the truth in nothing else, I shall at least be truthful in claiming that I am a liar. I think I can escape the censure of the world by my own admission that I am not telling a word of truth.'[31] The second part of his 'True Story' ends with a promise strikingly like Arbuthnot's last line, 'to continue in succeeding books'. 'The biggest lie of all', as the translator quotes a Greek scribe.[32]

There is a distinctly Lucianic turn to some of Samuel Butler's work. Among his comprehensive range of 'characters' are a number that bring satire and humour to bear on matters of philosophy and learning. Quite a few more pay testimony to the many forms that accusations of dishonety could take. The character of a libeller is one which seems to allude back to Lucian and Butler's image of an atheist is one of paradoxical contradiction. The specific depiction of a liar, in most respects derivative of Montaigne's essay and its imagery, ends on a promisingly paradoxical note. Should a liar tell the truth he becomes false to himself and deceives as many as when he lies.[33] This was written around the time Arbuthnot was born, though surviving only in manuscript was probably unknown to him. It is altogether less likely, however, that the whiggish John Dunton was as obscure. In his *Athenian Sport* he too fused satire and philosophy and did so persistently through a multivalent notion of paradox. Appropriately, it was a work that was not what it seemed, a few hundred paradoxes by an author who claimed to have garnered over one thousand. His first paradox, justifying his

still tediously copious miscellany is partly serious. The point of the
work, states Dunton, is to amuse and to stimulate thought. This
encouragement takes a broadly philosophic twist in the paradoxes
derived from recent science, for example, in the proposition that all
sciences should be reduced to one.[34] It takes a moral and satirical
turn by including as probability paradoxes essays on conventional
morality. Paradox 28 assures us that it is better to be chaste and
honest than loose and lying, the satire arising simply from the
presupposition that such a view is paradoxical in a morally corrupt
world. Another is an unacknowledged summary of Cornwallis essay
in praise of 'The Pocks'.[35] Although Dunton's title may seem to
take us to the philosophers of Athens and even to the paradoxes of
Socrates, the satiric style owes much more to Lucian and Menippus
in its acceptance of contradiction, its levity and its concern with
satirising forms of theory and intellectual pretension. Curiously
there is no properly formulated version of The Liar. The closest we
get is paradox 79 'In Praise of a Lye' and the one following, 'A Plot
and no Plot' which argues that 'when the Dissenters Plot to subvert
the Church of England, (in that very plot) they do their utmost to
serve and support it.'[36]

III

In 1712, however, appeared the anonymous and successful *The
Story of the St. Albans Ghost*.[37] In a discussion of authorship Koster
doubts, on the evidence of a computer analysis of its vocabulary,
whether it was in fact by Arbuthnot.[38] It may not be, but it has two
features that are instructively Arbuthnotian. First, it has the satiric
allegory form as would be used in the *John Bull* satires, largely
concerning Old Mother Haggie (Sarah Churchill); and more significantly, it handles the lying *topos* in a way not unlike Arbuthnot's
paradoxical treatment in *The Art of Political Lying*.

Given the standard pattern of accusation, it is not surprising that
dishonesty should be a subject of the satire. Yet the *St. Albans
Ghost* elaborates on the *topos* in part by encoding manifest implausibility and contradiction in its own claims about Mother Haggie.
It begins by warning the reader against superstition and falsehood,
indeed of papal traditions of false stories which may pass for truth,
standard fare no doubt, but views we have heard from Arbuthnot
before.[39] It then maintains that despite the prevalence of

superstition there are some stories which simply have to be believed as true. This is one, and the author follows this statement by providing marks of verisimilitude, thanking those who have been helpful in establishing its truth.[40] What, of course, is presented as true and carefully checked is manifest nonsense. Old Mother Haggie, we are assured is known to have ridden broomsticks; it is not known when she was born, it is therefore concluded that she was probably never born at all and since her correspondence with Old Nick is known 'beyond possibility of Disproof' no proof is offered.[41] Her powers were such that she was able to restore a maidenhead so many times that twenty men were satisfied of her daughter's virginity at marriage.[42] In *The Laws of Chance*, Arbuthnot had doubted that more than one in twenty women were virgins at marriage. Although at the level of political name-calling, the *St. Albans Ghost* is a Tory attack on the Churchills it is also a satire on superstitious rumours and the uncertainty of hearsay evidence. Like Lucian's 'Lover of Lies', it is a prime example of the very thing about which it warns at the outset, superstition and lies. Old Mother Haggie, though called a witch, is made more witch-like by the implausibility of the witch-hunt superstitions used to denigrate and defame her. There is a family resemblance between this subversive dimension to the *St. Albans Ghost* and what Arbuthnot would put into the world within a few months.

IV

Put most schematically *The Art of Political Lying* purports to tell us the truth about the inherent dishonesty of political discourse and by being a public political tract it is itself a political statement. Arbuthnot the politician says that all politicians are liars. We do not escape this structure by bracketing Arbuthnot from the author functions of the work. The Author of the 'Pseudologia' and the Projector, by his affirmative account, can be treated largely as one, the only clear point of difference being a matter of practicality not truth. Otherwise, they maintain that the political is a realm of dishonesty and the Projector's summary by being printed is in the political domain. The Author has invited public participation from politicians in perfecting the theory. Further what the Projector relates holds generally for the two principal forms of rule relevant to the political. If the rules of lying are constitutive of the political, then the account of the 'Pseudologia' as given by the Projector involves a falsehood if it claims to represent a truth. Conversely, if the 'Pseudologia' does not represent a truth, and it is proffered as a codification of all significant political truth, then it is a systematic lie. If, however, the rules are regulatory, a milder form of the paradox is still holds. What is useful and effective in politics is lying. The 'Pseudologia' is proclaimed to be useful and effective, a necessary and almost sufficient crib for the politician, in order to be so what it says is probably false. If this is so lying is not efficacious. In fact, the notion of rules in both senses is evoked in the text generally, and Arbuthnot, like his voices drew no real distinction between them.

In the most synoptic terms then, the whole pamphlet may be seen as a member of the liar family of semantic paradoxes. There is nonetheless, something distinctive about Arbuthnot's paradox. The members of the Liar family of which I am aware all rely on a specific semantic componant of the statement in order to produce a paradox. They consist of contradictions created directly by reflexivity, or in some more discursively expansive forms, consist of antinomies of vicious circularity. Resolution thus requires confronting the tensions of explicit semantic content. Thus, there is nothing paradoxical about Lucian's *Verae historiae* until the authorial voice change within the narrative that asserts that the only truth is that everything else within it is a lie. If Epimenides is not specified as a Cretan there is no semantic paradox in what he says about the Islanders. But as I have stressed, the voices of Arbuthnot's Author and Projector are fundamentally at one as to the truth content of the whole document. There is no inbuilt contradiction of the sort it would have been so easy to create within the text's semantic content. The paradox is made by the real author's act of producing the statement as a whole and placing it in the political world to which it refers. It is this that produces the elision of sense and reference typical of the Liar family. Arbuthnot's paradox may be called the entailed Liar. The paradox is entailed by our understanding the statement as a text. At the hermeneutic level, its significance consists in its drawing attention to the complex relationships between the notions of statement, context and agency. The paradox would suggest that although these may all formally be distinguished, they may not be separated as each requires implicit reference to the other two notions in some form. This sense of interrelatedness does not require anything as ambitious as a claim to have recaptured the mind of an historical author in order to establish a statement's meaning. We do not need here anything as politically specific as the hypothesis that Arbuthnot intended *The Art of Political Lying* to support Harley's political strategies, or anything as psychologically general as a motivation, his being moved by malice, a desire to reform, make money or amuse friends (these may in any case be more or less consistent with each other). We need do little more than to recognise a locus of agency (Arbuthnot/Morphew, or whoever) as doing something with a statement by placing it in the context to which that statement refers, in order to understand in turn how the contextualising act entails the statement's becoming paradoxical independently of any explicit formulation within the text itself.

Once recognised as an entailed paradox, a more politically particular significance arises from the work. The whole impetus of its reductivism is to force us to consider critically the nostrum that all politicians are liars and by virtue of this, the uncertain truth content of the political text itself. It becomes apparent that the text sustains the nature of the paradox by exemplifying the rules it describes. Through his personae Arbuthnot shows rather than asserts and thereby avoids the need to introduce any overt contradiction. He does so with such clear and consumate elegance that the reflexive application of his Author's rules could hardly have been inadvertent.

At the risk of over-reading, perhaps a clue to this is found in the opening simile of the mirror which has one further facet in this context. The claim is made that one side of the soul is like a mirror which reflects everything as it is. Strictly speaking, and one did not need Arbuthnot's knowledge of optics to realize the point, a mirror image shows things exactly as they are not; it is a reverse image and may forewarn the pedant at least of the tricks that Arbuthnot is about to play with truth. Be this as it may, towards the end of *The Art of Political Lying*, the Projector introduces us to the notion of the proof lie, that which a party puts about to test how far it can go. If a proof lie (such as transubstantiation) is swallowed, then everything else will be taken down as well. Put anachronistically, it is like the central myth at the heart of an ideology. The best example of a proof lie we have, however, is the cover of the pamphlet itself giving us the details of the book. As I have indicated, there is nothing in this to give the joke away, and if this 'swinger' is taken as genuine, then we have no reason to doubt the text that follows. It might also be noted that there seems to have been no precedent for Arbuthnot's fraudulent review, and so lacking obvious clues within the text, the reader was not able to assign the work to a genre or tradition of parodic satire. The advertisement and the title of the *magnum opus* and the Projector's word are all we can go on. As Thomas Morton had rhetorically asked of Robert Persons, '. . . why may not a lying title best befit the doctrine of *lying & dissimulation?*'[43]

Further, both advertisement and text follow the regulative rules concerning probability and verisimilitude upon which the 'Pseudologia' is so insistent. Lies should not be too far fetched, not too discrepant with what we think we already ·know. The clubs at which, we are told, the opus will be available, were all genuine and appropriately broad in their associations and clientele. The use of the authentic currency of rhetorical theory and plausible allusion to contemporary polities exemplify general advice. If the reader would know the Clubs to be places, the text exploits common places and the rhetoric that manipulated them. I have already suggested (ch. 5, III) that the text's relationship with the traditions of the *ars rhetorica* involves something of a 'prodigious lie'. The Projector's scepticism about one recommendation by the Author additionally has the effect of reinforcing the claim that he has perused the text with great care, and *a fortiori* that there is a text to peruse. Yet even insistence on his care in reading the volume we might doubt, for the book is summarized as having two chapter eights. Although this might be put down to Arbuthnot's legendary carelessness, he could count and the text was not corrected in the second edition. The 'Pseudologia' also insists that when lying, especially when countering other lies, appropriate touches of detail are vital. The proper response to the lie that the Pretender is in London is not an honest denial, but a claim (this gives the impression of deeper knowledge) that he got no further than Greenwich and turned back. Either way we end up with a correct conclusion: he is not there. Similarly with the lie about a treaty to introduce popery and slavery into England. The truth claim that there is no treaty is likely to be less effective in countering the assertion than the lie that some details in the treaty remain to be worked out – ironically a claim closer to the truth in 1712 than Arbuthnot *may* have realized. As the casuists might have asked, can specific falsehoods be made to serve a laudable end as well or better than truths? It is exactly this strategy of providing plausible detail that Arbuthnot adopts, so giving his

Projector a totally artificial authority, rhetorically speaking, an apposite *ethos*. The pamphlet seems to be a riposte to the pervasiveness of public dishonesty and it provides suitable subscription details, the quarto form of the book and its price. Sustaining the fiction to the end, the review closes with the promise of an account of the second volume. And twice *The Daily Courant* abridged these fictions in its advertising columns, providing a sustained context in which the false is nestled in with the true.[44] There is no direct evidence, but it is difficult to imagine the Scriblerians at least failing to see the full paradoxical nature of the satire. Yet by way of an epilogue, it is curious to note the play with contradiction in Pope's *Epistle to Dr Arbuthnot*. Ian Donaldson has drawn particular attention to this persistent structural feature of the work. It opens, he remarks, by asking for a closure, 'Shut the door good John' and it then exhorts Arbuthnot to lie for him, 'say I'm dead' in order that Pope can get on with telling the truth about satire – a distinctly casuistic move.[45] The result is to leave in parts an aura of indeterminacy. What, in the present context it may suggest is that Pope was alluding to or paying something of a compliment to his old friend whose love of paradox and satiric play with contradiction get their most concentrated expression in *The Art of Political Lying*.

What then, in rhetorical terms, is singularly decorous, a claimed truth presented through a lie, is also logically the most elegant elaboration of the paradox of the Liar; designed I suspect as such paradoxes always have been, to make us think about what is being said and what taken for granted and to examine the conditions under which a statement might be true or false. Because of this, *The Art of Political Lying* also functions as a solution to the problem of satiric reform in a fissiparous society. To be sure it is not a theory of satire, but it shows us what a satire might be like which avoids the repellent dangers of Pope's satiric nominalism on the one hand, the ethereal ineffectiveness of a fully realist satire on the other. Its own danger, as Swift realized at the time, was that it would pass human understanding.

NOTES

1   Niklas Luhmann, *Social Systems*, (1984) trans., John Bednarz Jr., with Dirk Baeker, (Stanford: University Press, 1995), chs. 11–12.

2   See Sister M. Geraldine, 'Erasmus and the Tradition of Paradox', *Studies in Philology*, 61 (1964), pp. 41ff; M. T. Jones-Davies, ed. *Le Paradoxe au temps de la Renaissance*, (Paris: Touzot, 1982), both of these studies are swamped by the diversity of materials in some sense paradoxical.

3   Plato, *The Republic*, trans. Paul Shorey, (Camb. Mass.: Harvard University Press, 1970 edn.) 508–9.

4   Cicero, *Paradoxa stoicorum*, trans. H. Rackman, (Harvard: Camb. Mass., 1942, 1960).

5   J. A. Barnes, *A Pack of Lies Towards a Sociology of Lying*, (Cambridge: University Press, 1994), p. 30.

6   Plato *Republic*, 415, A-B; 547, A-B.

7   Sir William Cornwallis, *Essayes or Encomiums*, (1616).

8   These first two senses of paradox are the ones, especially when given strongly religious overtones, discussed in Rosalie Colie, *Paradoxia Epidemica: The Renaissance Tradition of Paradox*, (Princeton: University Press, 1965).

9   R. M. Sainsbury, *Paradoxes*, (Cambridge: University Press, 1995 edn.), p. 1.

10 Sainsbury, *Ibid.*, pp. 127–9; cf. W. V. Quine, 'The Ways of Paradox', in *The Ways of Paradox and Other Essays*, (Harvard: Camb. Mass., 1976), p. 10ff with Graham Priest, 'The Structure of Paradoxes of Self-Reference', *Mind*, 103, (1994), 25ff.

11 Quine, 'The Ways of Paradox', p. 9.

12 Diogenes Laertius, *Lives of the Philosophers* II. 108, see C. L. Hamblin, Fallacies, (Methuen: London, 1970), ch. 3.

13 John Ibberson, *The Language of Decision*, (London: Macmillan, 1986), p. 15ff.

14 Philostratus, *The Life of Apollonius* (Camb.Mass.: Harvard University Press), bk. 5, 495.

15 Quine, 'The Ways of Paradox', p. 9.

16 Sainsbury, *Paradoxes*, p. 110.

17 Sainsbury, *Ibid.*, p. 145; Karl Popper gives a further version in an unusually charming dialogue on this whole issue: Theaetetus, '. . . the *Liar* . . . can be formulated by using indirect self-reference . . .' Socrates, 'Please give me this formulation at once'. Th. 'The next assertion I am going to make is a true one'. S. 'Don't you always speak the truth?' Th, 'The last assertion I made was untrue'. S 'So you wish to withdraw it? . . .' Th, 'You don't seem to realise what my last two assertions taken together amounted to'. See 'Self-Reference and Meaning in Ordinary Language' in *Conjectures and Refutations* (London: Routledge, 1972 edn.), p. 305.

18 John Allen Paulos, *I Think Therefore I Laugh*, (New York: Columbia University Press, 1985), p. 30 provides a modern version 'The Titl of This Section Contains Three Erors'.

19   Sainsbury, *Paradoxes*, stresses the point, but for a detailed discussion
     see F. Bottin, *Le Antinomie Semantiche nella Logica Medievale*,
     (Padua: Antenore, 1976); Aristotle, *On Sophistical Refutations*, trans.
     E. S. Foster (Camb. Mass.: Harvard University Press, 1955), e.g.
     172b, 11ff; 174b, 12ff.
20   Graham Priest 'Can Contradictions be True?' in *Supplementary Pro-
     ceedings of the Aristotelian Society*, 67 (1993), p. 35ff; *In Contradic-
     tion*, (Dordrecht: Nijhof, 1987); see also the brief discussion in
     Sainsbury, *Paradoxes*, ch. 6.
21   See for example, Jones-Davies, 'Avant-Propos', in *Le Paradoxe*.
     Paradoxes could instruct, amuse, stimulate or encourage devotion.
22   See Conal Condren, *George Lawson's Politica and the English Revol-
     ution*, (Cambridge: University Press, 1989), ch. 9.
23   John Donne, *Paradoxes and Problems*, ed. Helen Peters, (Oxford:
     University Press, 1980). In this way, the rationale for the paradox is
     much the same as Bacon's rationale for writing legal aphorisms, see
     above ch. 4, IV.
24   See Conal Condren, 'Cornwallis' Paradoxical Defence of Richard III:
     A Machiavellian Discourse on Morean Mythology?' *Moreana*, 24
     (1987), p. 5ff.
25   A. M. (Anthony Mundy), *A Defence of Contraries*, (1593), that one
     statement is addressed to the monarch the other to the reader, might
     complicate rather than resolve the apparent contradiction.
26   This adage, a variant of which was used by Nietzsche, might have
     stood as an epigraph for Paulos's *I think therefore I Laugh*. His
     starting point however, is just as pertinent, a comment of Wittgen-
     stein's that a good philosophical work could be comprised entirely of
     jokes, p. 5. On the philosophy of Sophistic tricks, see Hamblin,
     Fallacies, chs. 2, 3.
27   It is instructive that Mandeville's sub-title to *The Fable of the Bees*
     was a deliberate paradox which, he admitted, was chosen to entice the
     curious into his argument. For a discussion of this see, Phillip Harth,
     'The Satiric Purpose of *The Fable of the Bees*,' *Eighteenth Century
     Studies*, 2, 4, (1969), p. 327f.
28   Dryden, 'The Life of Lucian: A discourse of His Writings' (1696?
     1711), in *Of Dramatic Poesy*, 2, p. 211.
29   Lucian, *Works*, trans. A. M. Harmon, (Camb. Mass.: Harvard
     University Press, 1964), 3, p. 149.
30   Lucian, 'The Lover of Lies', in *Works*, 3, pp. 324–5.
31   Lucian, 'A True Story', in *Works*, 1, I.4. pp. 252–3.
32   Lucian, 'A True Story', in *Works*, 1, II.47, pp. 356–7.
33   Samuel Butler, *Characters* (1667?) ed. Charles Daves, (Cleveland:
     Case Western Reserve Press, 1970), 'A Libeller,' p. 172; 'an Atheist,'
     p. 163; 'A Liar,' p. 228–9.
34   Dunton, *Ibid.*, paradox 118.
35   Dunton, *Ibid.*, paradox. 75 'In Praise of the Clap.'
36   Dunton, *Ibid.*, paradox 80, p. 363.
37   Reprinted in Patricia Koster ed., *Arbuthnotiana*, Augustan Reprint
     Society, 154, (Los Angeles: University of California Press, 1972).

38   Koster 'Introduction', *Arbuthnotiana*, p. iv. It is suggested that Dr
      William Wagstaffe, who created *The Plain Dealer*, is more likely the
      author. The men were friends and collaboration is also possible. Swift,
      who denied authorship, knew the answer but never told.
39   *The Story of the St Albans Ghost* in Koster, *Arbuthnotiana* pp. 3–4; cf.
      The Usefulness of Mathematical Learning, pp. 9–10.
40   Koster, 'Introduction' *Arbuthnotiana*, p. v notes this feature; cf.
      Lucian, 'Slander', in *Works*, 1, p. 375.
41   *The St. Albans Ghost*, p. 6.
42   *Ibid.*, p. 8.
43   Thomas Morton, *A Full Satisfaction*, A4.
44   *The Daily Courant*, 3438, Oct. 18th; 3440, 21 Oct. (1712).
45   Ian Donaldson, 'Concealing and Revealing: Pope's Epistle to Dr
      Arbuthnot', *The Yearbook of English Studies*, 18 (1988), p. 181.

9 Conclusion

'Martin [said] that never having seen but one Lord Mayor, the Idea of that Lord Mayor always return'd to his mind; that he had great difficulty to abstract a Lord Mayor from his fur . . . On the other hand, Crambe, to shew himself the more penetrating genius, swore that he could frame a conception of a Lord Mayor not only without his Horse, Gown and Gold Chain, but even without Stature, Feature, Colour, Hand, Head, Feet, or any Body; which he suppos'd was the abstract of a Lord/Mayor. Cornelius told him he was a lying Rascal'.

John Arbuthnot and Alexander Pope *et. Al. Memoirs of the Extraordinary Life, Works and Discoveries of Martinus Scriblerus.* ch. 7, p. 120.

I

On his way to his own trial, Socrates met Euphythro who was having his father tried for impiety. For both men understanding impiety (*asēbeia*) was no trivial task. Euthyphro thought he understood impious actions – such as the abuse of a slave and the disrespectful treatment of his body; but asked Socrates, how was this possible without having a prior understanding of piety itself? A model (*parādeigma*) was needed through which the pious and impious might be securely classified.[1] How do we arrive at a general classification without first possessing some understanding of the specific actions we need to classify? This paradox of antecedents, Meno's paradox, has a number of manifestations, several of which have been threaded through this essay.[2] To one side of my direct concerns here, it arises in debates over the explanatory status of methodological individualism; and it is inherent in Dilthey's hermeneutic circle, that we understand the particular in terms of the general, the general in terms of the particular, text through context and context through text. And this has been an experiment in contextualization; an exploration and augmentation of changing meanings through shifting contexts. More directly, Meno's paradox is vital to the tensions between nominalism and realism; and to moral reasoning, as it is confined between the elucidation of guiding principles and the exploration of specific cases. Through these avenues, Meno's paradox has become important to understanding Arbuthnot's political satire. All these are hierarchical instances of a paradox of antecedents; that is, there seems to be a problem of what we need to establish first in order to proceed. Similar problems arise, however, when dealing with binary distinctions of the same semantic order such as good and bad, black and white true and false. A concept of one entails knowing simultaneously that it is not the other and that the other formally contradicts or contrasts with it.

II

There is a sense in which this whole species -of paradox is not as logically fraught as The Liar which demands assent to contradiction when we know that true and false are mutually exclusive predicates for the same statement. With the Meno family of paradoxes the difficulty arises because of the artificial assumption that in thinking through complex problems there has to be one fixed starting point. We begin it seems, with a *tabula rasa* and have to determine what we need as a condition for further understanding. It is exactly this which we do not have to accept. If we do resist on assumption of necessary priority, false dichotomies such as that between methodological individualism and holism, induction and deduction can be resolved. It was, I think, Popper who remarked, in Darwinian mood, that when it is asked what comes first, the chicken or the egg, the answer is a different sort of chicken (or vice versa).

One consequence of this experiment in contextualization is that the closer the textual reading, the more contexts are suggested. This point seems paradoxical only if we assume that text and context offer opposing epistemological categories and alternative strategies of interpretation; but they do not. The very notion of one presupposes some prior understanding of the other. That is, we always understand something when we begin and what may seem uncomfortably paradoxical really only attests to the interdependence of parts and wholes, generals and particulars, individuals and classes, texts and contexts. We understand one precisely because we have some rudimentary grasp of the other. One is the way of getting at the other. Thus rather than the individual, let alone the individual Lord Mayor, being opposed to some general classification, individuality is a function of classification, and classification the only means by which we can individuate.[3] In the same way, thick textual description and the concomitant etching in of multiple contexts ceases to look odd. It is for this reason also that we do not, cannot chose between opposing types of satire, general and particular, as Pope suggests we should, as we might chose to be Whigs rather than Tories. Something similar confronts would-be inductive scientists like Arbuthnot, who have perforce to rely on deductive reasoning – a point realized only too well by Arbuthnot's philosophical protégé Berkeley. And certainly, to bring us back directly to *Euthyphro*, moral reasoning requires a dialectic between principle and case. It was this recognition which made casuistry so powerful, yet when uncontrolled by principle, destructive.

III

Despite *The Art of Political Lying* being something of a complementary counterpoint to William Reeves' steadfast rejection of all casuistry, it is not quite the mirror image of Reeves in his pulpit, its paradoxical structure prohibits its being precisely that. Perhaps something of the tension between principle and case holds for the horizontal binary notions of truth and falsity, they are relational terms of the same order, each requiring the other before it can make sense; as Milton had urged in the *Areopagitica*, evil is in the world that we may understand good, *ergo*, lies that we may understand the truth.[4] Understanding, then involves a sort of hermeneutic oscillation between the two, captured symbolically in the Paradox of the Liar. The very specification of the political as a realm of lies presupposes some more truthful non-political domain by which we can measure deviance; but if they are so different, by what common standard does one judge either? Arbuthnot posits delineating realms of properly honest expectation, though by injecting his own truth to that effect into the political arena, he muddies the waters and must surely raise questions, whether intentionally or not, about how reliable such conventional dichotomies really are; from them come the easy accusations about political lying and the equally easy assumption of the high moral ground of censorious satire. The more we take the accusations at face value, the easier it is for us all to assume the status of satirist, to live in little more than a world of moral dogma. It is this which Arbuthnot's paradox unsettles if its philosophical dimension is taken seriously. Popper remarked, with David Hume in mind, that one of the great achievements of a philosopher was to see a riddle or paradox that had passed unnoticed. The perception 'disturbs laziness and complacency and arouses us from "dogmatic slumber".'[5] In this way, Arbuthnot's paradox of the political liar surely unsettles the satiric nostrum that all politicians are liars and so are the easy, routine victims of the satirist.

As Juvenal remarked, in a totally corrupt world, honest men can only write satire. How convenient, but if we are nudged towards satirical engagement with the world by taking simple truths for granted, there is nothing in this to encourage complaisance. As I have outlined in reference to Arbuthnot's satire, the self-promotional image of the satirist should not be accepted uncritically. Satirists trade conventionally, like all in politics, within the world of easy half-truth and are made vulnerable by the very moral indignation and pride that legitimates them. Achieving a morally and politically effective satire is what mattered back in the eighteenth century, something easily occluded by a latter day critical emphasis on the incidentals of literary quality.[6] As I have indicated, satire is as marked by complimentary modes of inneffectiveness as by forms of success. It is no accident that a time like the Stuart era, scarred by the eristic habits of moral passion was also a season of satire. *The Art of Political Lying* was one answer to the problem of being a satirist in what was seen to be a particularly naughty age. An antidote to enraged and dogmatic moralizing, it was an alternative to Pope's satiric nominalism and to Dryden's simplicity of theme and of direct and single precept.[7] It became a hostage to fortune because of its exquisitely self-indulgent cleverness, its gentle irony and more crucially its provoking indirection.

If we do think through and resist Arbuthnot's sidling rather than swaggering paradox, however, the one safe political conclusion to which it points, is that political discourse is never that simple, except perhaps for those be-limed in its prejudices and prearranged truths, mere satirists and satirists manqué. For all his distaste for the Whig war party and his personal loyalty to Queen Anne, Arbuthnot was not so enmeshed. Unconstrained by formal political office, he was even less trapped than his maverick friend Harley. If, as Lord Chesterfield remarked of him, Arbuthnot was a Jacobite by prejudice, a republican by reason, what else might we expect? To be sure he doesn't tell us this, but, and here he may well be following Dryden's injunctions, he does show us this conclusion if we accept the terms of his dilemma.[8] As *The Art of Political Lying* also shows, political discourse can be malleable and subtle, ambiguous and suggestive, misleading, as Reeves noted most explicitly, even when true; and in order to judge it we must in part understand its conventions even if we no not endorse them. This understanding Arbuthnot in his singular fashion half pretends to provide, while his account of his non-existent text conjures up less the spectre of Mother Haggie and her party affiliates brewing up a caldron of troubles, than the ghost of Gorgian rhetorical imperialism: the incantations of rhetoric make the political world as the spells of the magician re-fashion nature.[9] The analogy is useful but not sufficient. Nature does not need magic as politics needs language; but the success of the magician and the rhetorician is alike fugitive. The persistent Arbuthnotian image of the politician as artist is as fitting. Cards slip from the sleeve, paint fades. 'Adieu persuasive Eloquence' exclaims John Bull in *Lewis Baboon Turn'd Honest*.[10] So the political animal constantly seeks to reshape perception through the unreliable alambic of discourse, to marshall probabilities and to reformulate information concerning which John Bull and we, the judicious multitude, are caught between surfeit and starvation. Perhaps the vision of politics as a lie is only an '*artful Contrast, heighten'd with the Beauties of* Clar-Obscur' enabling us to maintain a face-saving distance from dangerous fools doing a necessary job. Lewis Baboon is usually honest enough if the job seems adequately done for the moment. Preparing for him to turn fully honest is a long wait for Godot. And if his transformation, as if by magic is one day unequivocal, '*we shall have* I know not what.'[11]

NOTES

1 Plato, *Euphythro*, 6D in *The Last Days of Scorates*, trans. Hugh Tredennick, (Harmondsworth: Penguin, 1969 edn.,) p. 26.
2 Plato, *Meno* ed. R. S. Bluck (Cambridge: University Press, 1961), 71e, 73c, pp. 154–156. See Bluck, 'Introduction' for the reduction of the issue to circularity, pp. 5–6. The Platonic doctrine of a process of recollection was the *deus ex machina*. The problems are implicitly in *Euthythro* and run through many of Plato's discussions of *forms*.
3 This is a Scriblerian point encapsulated in the crowd scene in *The Life of Brian*, when the crowd shouts in unison that they are all individuals and one man objects, 'I'm not'. Pure individuality is the empty rhetoric of the crowd. The classification individuality may suggest a version of Russell's paradox.

4   John Milton, *Areopagitica*, (1644) in *The Prose of John Milton*, ed. J.
     Max Patrick, (New York: Anchor, 1967), pp. 287, 297.
5   Karl Popper, 'On the Status of Science and of Metaphysics' in
     Conjectures and Refutations (London: Routledge, 1972 edn.), p. 184.
6   Gilbert Highet, *The Anatomy of Satire*, (Princeton: University Press,
     1964), Introduction.
7   Dryden, *The Original and Progress of Satire*, (1693) in *Of Dramatic
     Posey and Other Essays*, ed. S. Watson, (London: Dent, 1962), Vol. 2
     p. 146.
8   Dryden, *Ibid.*, p. 137.
9   Jacqueline De Romilly, *Rhetoric and Magic in Ancient Greece*, (Camb.
     Mass: Harvard University Press, 1975), pp. 11–20.
10   John Arbuthnot, *The History of John Bull*, eds., Alan W. Bower and
     Robert A. Erickson, (Oxford: Clarendon Press, 1976), p. 95.
11   *Ibid.*

Appendix A: A Scant Biographical Note on John Arbuthnot

The following involves no original work. It is merely a digest of what can be gleaned from Aitken's *Life*; the *DNB* entry; Beattie's *John Arbuthnot Mathematician and Satirist*; and Steensma's *Dr John Arbuthnot*.

The Arbuthnot family came from Kincardineshire in Scotland where John was born in 1667, one of three sons of Alexander an Episcopalian minister related to Lord Arbuthnot. John attended Marischal College, Aberdeen and was awarded an MA in 1685. At the Revolution of 1688–9 Alexander lost his living and retired to a small estate where he died in 1691. His acrimonious relationship with the Kirk resulted in a refusal to allow his family to set up a headstone on his grave. His sons, George, Robert and John variously left Scotland between their father's enforced retirement and his death. George, John's half-brother, joined the army, Robert went to France and became a successful banker. Each of these was to become a Jacobite, Robert doing much to finance the cause. John went to London around the time of his father's death and from the house of William Pate, a woollendraper earned money by giving mathematics lessons.

In 1692 he published *The Laws of Chance*, his adaptive translation of Huygens *De rationiis in ludo alea*. This may have been the stimulus for him to do his own work on probability theory during the 1690s and which was to surface as his *Argument for Divine Providence* (1710). In the meantime, in 1694 he went up to University College Oxford, coming under the wing of Dr Arthur Charlett, Master of University College; but it was from St Andrews that he took his degree in medicine in 1696. From then he maintained an increasingly successful practice in London. Apart from being close friend and physician to Pope and Gay, and Queen Anne, his practice would include Lord Chesterfield, Lady Suffolk, and the playwright William Congreve. While his practice was developing, he was establishing a reputation as a serious scientific figure by publishing *An Examination of Dr Woodward's Account of the Deluge* (1697) and *An Essay on the Usefulness of Mathematical Learning* (1701). He was elected to The Royal Society in 1704.

In the following year he was fortunate to be on hand when Prince George was taken ill and no doubt partly as a reward for help and appropriate prescription, was appointed Physician Extraordinary to the Queen. In the same year he made his first foray into antiquarian studies by publishing *Tables of the Grecian, Roman and Jewish Measures*, and was appointed by the Royal Society to mediate between Newton, Halley and Flamstead over the publication of Flamstead's astronomical tables. He did so without conspicuous success. The following year *A Sermon Preach'd to the People at Mercat Cross* provides evidence of his emerging status as an informal adviser on Scottish affairs. In 1709 he was made Physician in Ordinary to the Queen and was by this time probably a significant councillor. During the next year he became a Fellow of the College of Physicians and delivered his famous address before the Royal Society on Child Mortality (*An Argument for Divine Providence*). In 1711 he was given a sinecure in the Customs. By this time he had met and become a close friend of Swift, with whom he formed 'The Brothers' Club and of Robert Harley the Queen's chief minister. During a year which saw the most intense political turmoil around Harley's ministry, Arbuthnot was appointed by The Royal Society to the committee set up to determine the rival claims of Newton and Leibniz as to the invention of the infinitessimal calculus, the 'theory of fluxations'. It was in the course of 1712 that he wrote his major satires, the *John Bull* pamphlets and *The Art of Political Lying*.

In 1713 he was appointed Physician at the Chelsea hospital, and at the end of it, or in early 1714 the Swift/Arbuthnot circle from The Brothers Club amalgamated with the younger Pope and Gay to form the Scriblerus Club (see Appendix B). Political tensions were destroying the alliance between Harley (now Lord Oxford) and Bolingbroke, although Arbuthnot was to remain on good terms with both men. The death of the Queen in August transformed Arbuthnot's situation. The Scriblerus Club broke up, he was deprived of his offices at the Hanoverian Succession, although he seems only to have been politically persona non grata because of his Tory associations in a new Whig world. He was, however, by no means socially ostracized. For a short time following the Queen's death, he went to France, presumably to his brother Robert. There is, *so far*, no evidence that he was implicated in the Jacobite Rising of 1715 but I suspect he toyed with involvement; he remained on intimate terms with Robert who was significantly committed. Before then, John had returned to London to carry on with his medical practice. In 1716 he printed the *Humble Petition of the Colliers*, and in the following year had a hand in the notorious play *Three Hours After Marriage*. He went to France again in 1718. In 1719 he participated in the controversy over the efficacy of small pox inoculation using statistical evidence to support the procedure. In 1722 he published *Annus mirabilis*. In the following year he was appointed Second Censor in the College of Physicians. After that he had published the *Quidnuncki's* (?) and *The Reasons Humbly Offer'd*. In 1727 he returned briefly to his antiquarian interests publishing *Tables of the Ancient Coins* and he also delivered the Harveian Oration at the College of Physicians. It was around this time that Arbuthnot's joint Scriblerian project with Pope, *Peri Bathous* broke down (see Appendix B) but to no lasting detriment to their on-going friendship. The Motte Miscellanies were published including both *Peri Bathous* and *The Art of Political Lying*.

In 1729 Arbuthnot's *Virgilius Restauratus* was included with the valiorum edition of Pope's *Dunciad*. Over the next two years Arbuthnot's wife Margaret died, and then his son Charles. During 1731 Arbuthnot produced the *Brief Account of Mr. John Ginlicutt's Treatise* and his *Essay concerning the nature of Aliments*. In 1733 he wrote or finished the Scriblerian *Essay Concerning the Origin of Sciences*, and the scathing *Epitaph on Francis*

Charteris. John Gay died. Despite increasing bad health brought about by too great a love of good food and company, Arbuthnot carried on working. In 1734 he moved from London to Hampstead for an improvement in air and published *An Essay of the Effects of Air*, and the next year his final work, *Gnothi Seauton*. Throughout this time he had kept up an impressive correspondence on scientific and other matters and had remained in touch with his close friends Swift in Ireland and Pope much nearer to hand. Pope's *Epistle to Dr Arbuthnot* was probably polished to appear before his friend's death. Arbuthnot died partly from complications of 'the stone' in pain but peacefully in 1735 and was survived by two daughters, Margaret and Anne and by his son George who would be one of Pope's executors.

George would also cast doubt on the authenticity of the works that came out under Arbuthnot's name, but his sweeping dismissal of so much has only complicated issues of authorship. The works published as Arbuthnot's attest to his reputation as a writer. Initially that reputation was helped by the high profiles of his friends and by the received image of his character. This was of a man of great charm and conviviality, and of brilliant conversation in the old broad sense of the term; one who could make and keep friends with ease because of a selfless generosity and willingness to put them first. As a physician he would treat those who could not pay him with as much diligence and sympathy as those who could. Singling him out for his learning, character and piety, Dr Johnson reproved only his gluttony. Whatever the reason for his widely attested sociability and consideration for others, Arbuthnot could make and keep an extraordinary diversity of friends even despite their often destructive rivalries, tensions and vanities. A veritable Dr Omnium at one with Swift, Pope, Gay, Parnell, Handel, Harley, Bolingbroke, Queen Anne, even Isaac Newton, a man whose capacity for friendship seems to have been withered at birth. An appearance of equanimity was no doubt part of it. Swift wrote

'How those I love my death Lament.
Poor Pope will grieve a month, and Gay
A week, and Arbuthnot a day.'[1]

If such lines show more than the requirements of scansion, they might suggest the courtier's emotional opacity as much as a resilient cheerfulness in explaining Arbuthnot's persistent and admired public demeanour. We'll never know and *The Art of Political Lying* which might have welled from either source, will never help us.

NOTES

1   'On the Death of Dr Swift' (1731) in *The Works of Jonathan Swift*, ed. Thomas Sheridan and John Nichols, (London, 1808), vol. 17, p. 190.

Appendix B: The Art of Political Lying, Scriblerian Satire and the Authorship of *Peri Bathous*

Throughout the argument I have made passing comment on *The Art of Political Lying's* relationship to the Scriblerus Club. Because the club was formed after the work's publication, it was not appropriate to discuss it at any length. I want now, however, to provide a more sustained comment on the Scriblerians, for if the Club cannot strictly be seen as a context for *The Art of Political Lying*, the reverse most certainly holds; and in particular, holds for what was perhaps the last of the major Scriblerian satires, *Peri Bathous, or The Art of Sinking in Poetry*.

The Scriblerus Club was itself was the amalgamation of a group, including Arbuthnot, that gathered around Swift, and one associated with the younger Alexander Pope. It was formed under the initial leadership of Swift around the beginning of 1714. Then, according to Swift himself, Arbuthnot became the seminal force, after Swift's departure from London in June, and it was in Arbuthnot's rooms the Club usually met from its inception on a weekly basis until Queen Anne's death in August 1714. The principal members were Swift, Pope, Arbuthnot, Parnell, Atterbury, Gay and Harley and with the printer John Morphew sometimes present.[1] From what I have argued, it will be clear that Scriblerian satire was neither ad hoc, nor is it adequately understood in terms of a blanket hostility to the modern world or as being directed against intellectual pretension. As Brean Hammond has emphasized, although such characterizations are broadly true, the Scriblerians more precisely shared Swift's deep-seated worries about language and were highly suspicious of science and modern learning because of a fear that the world would be seen only in terms of its material constituents. At once such reductive materialism would leave no room for the spiritual and inspirational, and it would allow the illusion to flourish that knowledge was nothing but the mechanical application of rules to data.[2] As Atillio Brilli has argued at greater length, Scriblerian satire was under-pinned by a genuine fear of science and was an attempt to control its social and spiritual effects.[3] Concomitantly, there was a fear of the language of science and learning channelled through contemporary evocations of the traditional suspicion of rhetoric and exemplified most obviously in Locke's *Essay*.[4] *Homo mechanicus*, to use Hammond's expression, in all his manifestations is the target.[5] There is, argues Hammond, a good case for seeing the Club's satire as sufficiently general and coherent for the term Scriblerian to be extended beyond its core referents, the works of Martinus

Scriblerus, as one might extend a satiric classifier such as Menippean.[6] Not withstanding the dangers of applying such labels generously, Hammond's point immediately suggests that Arbuthnot's seminal role in the Club must be accepted at face value; for, of its members, he was the only one really able to sort scientific wheat from chaff, as much philosopher as mathematician, the only one who was as engaged with modern science as he was imbued with ancient learning. The Scriblerian Club was a site for his life-long pursuit of sound science and mathematics by other means. More specifically, it requires only a minor chronological adjustment carrying the barest taint of anachronism to classify *The Art of Political Lying* as a piece of Scriblerian satire. All the general features of Scriblerian satire can be teased out from the differing contexts in which I have placed it. As I have noted in the main text, (ch. 4, I) Swift's initial characterization of the pamphlet when writing to 'Stella' (9th Oct. 1712) was as a satire of learning rather than of politics. It is odd that both Brilli and Hammond quite overlook the most succinct exemplification of their arguments.

The Club got its name from the central satiric creation, the epitome of *homo mechanicus*, Martinus Scriblerus, for whom the plan was to provide a biography, to help explain his manic and unreflective penchant for misplaced learning, rule- fetishism and vulgar reductionism; a series of works by him triumphantly displaying these faults, and with consummate cheek, to claim on his behalf any genuine work which seemed to the Club members sufficiently 'Scriblerian'.[7] This was the most elaborate of satiric 'projects'. In fact, we have the biography and some work in logic, metaphysics and medicine in *The Memoirs of Martinus Scriblerus*, and *Peri Bathous*, the last word on poetry and rhetoric. It may seem churlish, but for all the definitive nature of what Martinus did write, the quantity of his output is disappointing. Nevertheless, a catalogue of the works which involved cooperation between members of the Club, or which require a Scriblerian context to make full sense of their point, would include *The Beggars' Opera*, *Three Hours After Marriage*, *The Dunciad* and *Gulliver's Travels*.[8] In the last of these, for example, the 'Voyage to Laputa' has a distinctly Scriblerian flavour perhaps even an Arbuthnotian one in its satire of scientific theories and in its elaborately reifying imagery of the reference functions of language.

From what little evidence survives, the Scriblerians operated like a highly social script-writing club, organizations for which it is not only often impossible but misleading to assign determinant degrees of responsibility. There is a dubious ease with which, for example, *Three Hours After Marriage* is attributed to one person, John Gay who took responsibility for it. Contemporary attribution, however, was to the 'Triumvir' of Gay, Arbuthnot and Pope.[9] The mocking of Dr Woodward's theories through the character Dr Fossile are at one with Arbuthnot's early demolition of them, but such theories were well in the public domain by 1717 and had done their work, so Gay might have used them without any suggestion or creative in-put from his friend. As Fossile is a medical man, Arbuthnot may have given help or advice on medical terms and jokes. More suggestive is the paradoxical reflexivity of the play, a satire on comedy whose dénouement involves happily providing one of the characters with a plot for a comedy. But despite his known involvement there is nothing more that might be disengaged as Arbuthnot's contribution; and to go this far is to privilege guess-work over evidence. More to the point, there is nothing that reference to *The Art of Political Lying* might elucidate. It may be tantalizing but sensible to see *Three Hours* as a Scriblerian piece principally by Gay, with the classifier Scriblerian doing serious predicative work.

The situation with respect to *Peri Bathous*, however, I want to suggest is different. It cannot properly be predicated Scriblerian or contextualized without relating it to *The Art of Political Lying*. *Peri Bathous, Or the Art of Sinking in Poetry* was produced as the work of Martinus in 1727. The text appeared in a volume of the *Miscellanies* of their own work and some of that of their friends prepared by Pope and Swift.[10] The satire is principally attributed to Pope with some assistance by Arbuthnot and possibly other Scriblerians.[11] Its themes are typically Scriblerian, of misapplied learning, reductionism, and suspicion about rhetoric and its misuse. Its use of an ancient model to attack modern writing is broadly Swiftian, yet another skirmish in the Battle of the Books. In her careful 'Introduction' to *Peri Bathous*, E. L Steeves is representative of the received view.[12] She accepts that one cannot be definitive about the work's authorship but writes as if regardless, it is really by Pope.[13] In this I believe she is both right, we cannot assign certain authorship, and wrong, we need to accept the consequences of that fact.

The grounds for having to accept shared authorship are clear enough. The pseudonym Martinus Scriblerus, was, as it were, the Club's signature. I think however, that we can, cautiously go further. There seem to be traces of a rough division of labour between Arbuthnot and Pope. Apparently, *Peri Bathous* was begun in the first seminal period of Scriblerian energy, around 1714.[14] But after the Scriblerian diaspora Arbuthnot and Pope took the idea forward on their own. The only clear evidence of authorial responsibility comes from Pope. Apparently, he got impatient with Arbuthnot's procrastination, arising from a preoccupation with some other work. Pope took it out of his hands, 'methodized' and polished it and put it into print. He was certainly very pleased with the work stating, correctly, that it could be read as a straight *ars rhetorica*, and claiming it largely as his own.[15] After Pope's self-satisfaction opinions have divided sharply on the quality of the finished product, invariably with unstated criteria of appraisal making it unclear exactly what is being asserted. At any rate, Aitken and Beattie both thought little of *The Art of Sinking* and possibly because of this, were reluctant to see much of Arbuthnot in it; Steeves and Ross have been more impressed, but as I have noted, have thought it safer in the bosom of Pope. Increasing interest in rhetoric and greater self-awareness about the difficulties of passing literary judgements may well help remedy the work's neglect.[16] Be this as it may, Pope later revised his claim giving Arbuthnot a greater share of responsibility.[17] None of this tells us very much, even assuming Pope is scrupulously reliable. We do not know the state of the project after formulation by the Scriblerian Club, or in what condition Pope took it from Arbuthnot. To Methodize had meant to put in a schematic picture, a family tree of conceptual relationships in the idiom of Peter Ramus, but more broadly could have meant to organize in terms of simple sub-headings, and the process of polishing can mean almost anything. What we have is surviving evidence of co-authorship and we should accept it for what it is and take seriously, after Hammond's suggestion, the implications of pseudonymous authorship.

Notwithstanding this, however, *The Art of Political Lying* has been almost totally ignored as a context, or pre- text for the *Peri Bathous*. Kerby-Miller remarks briefly that it may be part of the literary context for later Scriblerian output, and Bruneteau has gone further calling it the prelude to the activities of the club.[18] It is this suggestion I wish to take up. If *Peri Bathous* was originally conceived in the first phase of Scriblerian activity, then it does not post-date *The Art of Political Lying* by more than a year or so, and even in its final Popised manifestation, the similarities between the works are, *in toto* striking. The titles have the same form; Greek with English sub-title, epitomizing an ancient modern dichotomy. Whereas *The Art of Political Lying* is an allusive pastiche of the claims of the *ars rhetorica*, *Peri Bathous* (*On the Deep*) is for the most part a parody of Longinus' *Peri Huplous*, (*On The Sublime*). Yet, in another sense the work is, like *The Art of Political Lying*, deceptively wide ranging, for it brings together traditions of poetic as well as rhetorical theory. In each work the central *topos*, the love of lying and the love of bathos respectively, is seen as implanted in and stemming from the soul.[19] In each case then, there is a spiritual explanation for a mechanical reductionism. The author of *The Art of Political Lying* is anonymous, but he makes the same sort of claim as does Martinus Scriblerus: he is codifying what has lain only in confusion, in each case, a proposition made doubly absurd by being an inversion of the very familiar, a long- standing tradition of rhetorical rule books mediating ancient wisdom for the modern world.[20] Each work is presented as being systematic and complete and of the greatest practical utility. Each in short, is the very sort of vulgar teaching crib to which the Scriblerians objected, not least in the gleefully shared propensity to reduce everything to a learnable Rule.[21] Both works proceed by a systematic inversion of what can be taken seriously; and as is so appropriate to mechanical rule-mongering, each suggests the establishment of an over-seeing society.[22] As E. L Steeves remarks of *Peri Bathous*, it is a 'sparkling piece of inverted argument', written with the air of 'bland detachment' suggestive of John Arbuthnot.[23] She goes on to remark that it is difficult to tell whether the purpose of *Peri-Bathous* is more to satirise or to set out the principles of serious art; she might even better have been commenting on *The Art of Political Lying*.[24] This problem arises, not simply because there is always a serious point to satire, but because of each work's studied reductionism. As Hammond has correctly stressed, Scriblerian satire was typically reductionist in taking something in itself of value and pushing it to an absurd extreme.[25] This is why Martinus although foolish, is neither stupid nor uneducated, a man as Pope styled him of 'capacity enough';[26] it is also why we must see the Author of the 'Pseudologia' as more than a mere butt for Arbuthnot's humour. There is truth in what he is saying; it is the reader's job to think through and find out how much and where. Hence the aptness of Pope's remark that *Peri Bathous* could be read straight as an *ars rhetorica*. Each work also exploits the tendency to reify intellectual constructs by turning words into material objects – it is typical also of Swift's account of the Academicians of Lagado in Gulliver's voyage to Laputa, some of whom use objects for words while others work with a mechanical language box. The paradiastolic effect is to create an image of a merely material world. This is most clear in the Appendix of *Peri Bathous* with its elaboration of the rhetorical tool box with its drawers for the *genera* of rhetoric, epideictic, forensic and deliberative, sub-divided into cells in which the ingredients for poetry can be kept. It is an invention in no way suggested by the parodic source. Steeves suspects an Arbuthnotian 'influence'.[27] Well she might, given the language of *The Art of Political Lying*, but the term influence is particularly misleading here, presupposing a solid, basically Popian identity which she admits cannot be shown.

There are further incidental features *Peri Bathous* shares with *The Art of Political Lying*. In one we have 'velocity' of lies, in the other 'velocity' of poetic judgement.[28] In both works there is play with optical imagery.[29] *Peri Bathous* also exhibits an interest in the deceptions and transformative qualities of public language, with the powers of *paradiastole*. Chapter Fourteen includes advice on how to change a vicious man into a hero and by the Rule of transformation, change vices into virtues – a typically hallowed fear about the dangers of rhetoric, so calmly exploited in *The Art of Political Lying*. Cowardice, *Peri Bathous* informs us may be metamorphosed into prudence, corruption into patriotism. In short, as in *The Art of Political Lying*, rhetorical redescription is elided with creation, the confecting of a spurious reality (see above, ch. 7, V). By the Rule of Contraries, the less virtue a man has the more he needs it, a simple descant upon the strategies of Arbuthnot's additory lying.[30]

Taken individually, the textual similarities I have noted would establish virtually nothing, but together they amount to more than a series of *ad hoc* overlaps, coincidences or 'influences'. They relate both to the general shape and intellectual structure of *Peri Bathous* and to incidental detail but not to examples or to textual organization. This is to hazard that Arbuthnot with the other Scriblerians initially refined the masterplan for the work distantly out of Pope's suggestion for a collection of the works of the unlearned, and the Swiftian satiric agenda established by *The Tale of a Tub*; but more immediately and precisely they did so out of *The Art of Political Lying*, the anonymous Author of one, as it were, being born as Martinus Scriblerus of the other. Kerby- Miller perceptively remarks that Martinus Scriblerus is a sort of Quixote who usually needs a foil in the form of a servant to reveal his bizarre form of learned ignorance; this is close to the relationship between the Author and Projector of *The Art of Political Lying*, almost an anonymous Martinus and his devoted helper.[31] The rich (and wonderfully upsetting) exemplary detail in *Peri Bathous*, providing in *Private Eye* terms, a systematized 'Pseuds Corner', would take more time, and obviously once supplied and used leave no trace of its collectors.

The Scriblerian group broke up, meeting only occasionally after the death of the Queen; Arbuthnot and Pope then took up the idea again and Pope took it over. Arbuthnot may have lost interest or have been working on something else c.1726–7, but in the main body of the argument, (chapter 3, VI), I have suggested that a tendency to differ on issues of satiric strategy, inadequately captured in the distinction between general and particular, also needs to be taken into account. It was believed relevant at the time.[32] As I have intimated above (ch. 3, VI) it may even have been the embarrassments attendant upon their collaboration in *Three Hours* that cast a lingering shadow over the consequences of overt co-operation. Indeed, it seems to me that the greatest point of difference between *The Art of Political Lying* and *Peri Bathous* is that where the rhetorical principles of lying are only hypothetically and generally illustrated in the former, the latter is deliciously replete with quotations and the initials of their hapless authors. It thus provided something of what people feared would come from the pen of Pope when he was thought to be responsible for *Three Hours After Marriage*, an attack on all modern poets. This specificity seems most likely to have been the imprint of Pope, who, I guess, selected, arranged, possibly from quite a wealth of Scriblerian gatherings, and no doubt polished the style. Punctuation and spelling never had more than a fugitive grasp on Arbuthnot's imagination. If this is close to the mark, Pope's main attack on writing then, remains *The Dunciad*, he did just enough in *Peri Bathous* to spur him on, its reception prepared the ground wonderfully for him.[33]

Much of this is hypothetical and is not designed to replace Pope with Arbuthnot as author. But I believe we can say that when considered in the context of *The Art of Political Lying*, *Peri Bathous* is cast in a sufficiently different light for us to be fastidious in keeping to dual or multiple attribution. It is not only most safely but most properly seen as a Scriblerian text, and in the full sense of that term. Unfortunately, it has been more convenient to assign texts to single names, and if we want to encourage their study, the more famous the name the better. If the author is dead, it doesn't matter. This is an ironic epitaph to Swift's and Pope's hostility to what was foisted upon them and about which they so angrily complained in the 'Preface' to those volumes in which *Peri Bathous* and the second version of *The Art of Political Lying* appeared together.

NOTES

1   See Charles Kerby-Miller, ed. *The Memoirs of the Extraordinary Life, Works and Discoveries of Martinus Scriblerus*, (Oxford: University Press, 1988), 'Introduction' for a fine overview of the Scriblerians, esp. pp. 2–23 for the process of formation; see also Claude Bruneteau, 'John Arbuthnot (1667–1735) et les idées au debut du dix huitieme siècle', Doctoral thesis, Université de Lille (1974), vol. 1, pt. 1, ch. 2, pp. 59–61; Joseph Levine, *Dr Woodward's Shield: History, Science and Satire in Augustan England*, (Ithaca: Cornell University Press, 1977 and 1991), ch. 13, and p. 239. The presence of Robert Harley is a striking exemplification of the alliance between politics and letters, and in twentieth-century terms an improbable one, like Harold Macmillan co-writing for 'Beyond the Fringe', or Margaret Thatcher being a Python.

2   Brean Hammond, 'Scriblerian Self-Fashioning', *The Year Book of English Studies*, 18, (1988), p. 108ff.

3   Attilio Brilli, *Retorica della satira con il Peri Bathous, o L'arte d'inchinarsi in poesia di Martinus Scriblerus*, (Bologna: Il Mulino, 1973), p. 37ff, 50–6.

4   Brilli, *Ibid.*, p. 60ff.

5   Hammond, 'Scriblerian Self-Fashioning', p. 118.

6   Hammond, *Ibid.*

7   Kerby-Miller, 'Introduction', p. 29ff.

8   Kerby-Miller, *Ibid.*, p. ix.

9   *A Key to the New Comedy*, (1717); See John Harrington Smith, *Three Hours After Marriage*, Augustan Reprint Society, 91–92, 'Introduction', p. 1. The Royal Shakespeare Company's revival of the play in 1996 attributes the work to all three friends.

10   Jonathan Swift and Alexander Pope et al., *Miscellanies in Prose and Verse* (1727) vol. 3.

11   George Aitken, *Life and Works of John Arbuthnot*, (Oxford: Clarendon Press, 1892), p. 331 excludes it on this basis; Alexander Ross, 'The Correspondence of John Arbuthnot', Cambridge University unpublished PhD thesis (1956), vol. 1, p. 67–8, can find little sign of Arbuthnot's involvement; but see the discussion by Lester M. Beattie, *John Arbuthnot, Mathematician and Satirist* (Camb. Mass.: Harvard University Press, 1935 and New York: Russell and Russell, 1967), p. 278ff.

12   *The Art of Sinking in Poetry: Martinus Scriblerus' Peri Bathous, A Critical Edition*, ed. E. L. Steeves, (New York: Columbia University, The Crown Press, 1952). References to *Peri Bathous* are to this edition.

13   Steeves, 'Introduction' p. xxxii and passim; see also, by implication Kerby-Miller, 'Introduction', p. 55; Alvin Kernan, *The Plot of Satire* (Yale: New Haven, 1965), ch. 2.

14   Kerby-Miller, *Ibid.*, pp. 54–6.

15   Steeves, 'Introduction', p. xxiv–xxv; Kerby-Miller, 'Introduction', pp. 54–5. Kernan, *The Plot of Satire*, notes Pope's remark, tying it plausibly to a *topos* in Quintilian, *Institutio oratoria*, ed. and trans. H. E. Butler, (Camb. Mass.: Harvard University Press, 1920), 8.3.58, style may be corrupted in as many ways as it may be embellished/armed (*ornatur*). He then proceeds to use it as a key to the master tropes of satire, p. 32ff.

16   See for example Brilli, *Retorica della satira*, at length.

17   Steeves, 'Introduction', p. xxvi.

18   Kerby-Miller, 'Introduction', p. 72; Bruneteau, 'John Arbuthnot et les idées' pt. 1, ch. 2, pp. 57, 59–61.

19   *Peri Bathous*, ch. 2; cf. *APL*, p. 7.

20   *Peri Bathous*, ch. 1, pp. 5–7; cf. *APL*, p. 3 (abridgement of ch. 1).

21   *Peri Bathous*, ch. 3 on the necessity and usefulness of rules; ch. 4 on the rules of art as a guide to 'the deep' and chs. 4, 5, 14 and ch. 15 is a recipe for an epic poem.

22   *APL*, p. 17; *Peri Bathous*, ch. 14.

23   Steeves, 'Introduction', pp. xxx, xxxi.
24   Steeves, *Ibid.*, liii.
25   Hammond, 'Scriblerian Self-Fashioning, p. 114.
26   Kerby-Miller, 'Introduction', p. 30 stresses the significance of this.
27   Steeves, 'Introduction', p. 186.
28   *Peri Bathous*, pp. 5–6.
29   *APL*, p. 7; *Peri Bathous* ch. 5, p. 18, in each case the imagery is used to denigrate the subject of the satire.
30   *Peri Bathous*, p. 79.
31   Kerby-Miller 'Introduction', p. 30.
32   Leonard Welsted and James Moore-Smythe, *One Epistle to Mr. A Pope*, (1730), pp. vi–vii, see Steeves, 'Introduction' p. xxvii; also Ross, 'Correspondence', 1, 67–8; Bruneteau, 'John Arbuthnot et les idées', Appendice C for an even greater emphasis on the satiric differences between Pope and Arbuthnot.
33   Steeves, *Ibid.*, shrewdly suggests that Pope brought out *Peri Bathous* in part to flush out potential victims of *The Dunciad*, she calculates that about half of the satirized authors are common to both works, pp. xlvi–xlvii, l–li.

Appendix C: The Text of
The Art of Political Lying

I have followed capitalization, spelling and punctuation from the 1712 edition, but the long s has been modernised. The matter in square brackets indicates alterations in the second edition, or transliterations. The matter within pointed parentheses ({ . . . })marks simple omissions from the 1727 edition. The page numbers in square brackets are to those of the 1712 edition.

PROPOSALS

For PRINTING
**A very Curious Discourse, in
Two Volumes in *Quarto*,**

Intitled,

ΥΕΥΔΟΛΟΓ ΙΑ' ΠΟΛΙΤΙΚΗ' ;

OR, A

TREATISE of the ART
of

PoliticalLying
WITH

**An Abstract of the First Volume
of the said TREATISE.**

LONDON:
Printed for *John Morphew*, near *Stationers-
Hall.* 1712. Price 3d.

[5]

There is now in the Press a Curious Piece, intitled [Pseudologia Politicē]; or, *A Treatise of the Art of* PoliticalLying. Consisting of Two Volumes in 4 to.

The Propsoal are,
I. *That if the Author meets with suitable Encouragement, he intends to deliver the First Volume to the Subscribers by* Hilary-Term *next.*
II. *The Price of both Volumes will be, to the Subscribers, Fourteen Shillings; Seven whereof are to be paid down, and the other Seven at the Delivery of the Second Volume.*

III. *Those that Subscribe for Six shall have a seventh gratis; which reduces the Price to less than Six Shillings a Volume.*
IV. *That the Subscribers shall have their Names and Places of Abode Printed at length.*
Subscriptions are taken in at *St. James's* Coffee-house, *Young Man's* at *Charing Cross*, the *Grécian*, *Brydges's* by the *Royal Exchange*, and most other Coffee- houses in Town.[1]

[6]

For the Encouragement of so useful a Work, it is thought fit the Publick should be inform'd of the Contents of the First Volume, by one who with great Care perus'd the Manuscript.

The Author, in his Preface, makes some very judicious Reflexions upon the Original of Arts and Sciences; That at first they consist of scatter'd Theorems and Practices, which are handed about amongst the Masters, and only reveal'd to the *Filii Artis*, till such time as some great Genius appears, who Collects these disjointed Propositions, and reduces them into a regular System. That this is the Case of that Noble and Useful Art of *Political Lying*, which in this last Age having been enrich'd with several new Discoveries, ought not to lie any longer in Rubbish and Confusion, but may justly claim a Place in the *Encyclopaedia*, especially such as serves for a Model of Education for an able Politician; that he proposes to himself no small Stock of Fame in future Ages, in being the first who has undertaken this Design; and for the same Reason he hopes the Imperfection of his Work will be excused. He invites all Persons who have any Talent that way, or any new Discovery, to communicate their Thoughts, assuring them that honorable mention will be made of them in his Work.

[7]

The First Volume consists of Eleven Chapters.

In the first Chapter of his excellent Treatise, he reasons Philosophically concerning the Nature of the Soul of Man, and those Qualities which render it susceptible of Lyes. He supposes the Soul to be of the Nature of a *Plano-Cylindrical Speculum*, or Looking-Glass; that the plain side was made by God Almighty, but that the Devil afterwards wrought the other side into a Cylindrical Figure. The plain side represents Objects just as they are, and the Cylindrical side, by the Rules of Catoptricks, must needs represent true Objects false, and false Objects true, but the Cylindircal side being much the larger Surface, takes in a greater Compass of visual Rays. That upon the Cylindrical side of the Soul of Man depends the whole Art and Success of *Political Lying*. The Author in this Chapter, proceeds to reason upon the other Qualities of the Mind; As, great fondness of the *Malicious* and the *Miraculous*: The tendency of the Soul towards the *Malicious*, springs from Self-Love, or a Pleasure to find Mankind more wicked, base, or unfortunate, than our selves. The Design of the

Miraculous, proceeds from the Inactivity of the Soul, or its Incapacity to be moved or delighted with anything that is vulgar or common. The Author having establish'd the Qualities of the

[8]

Mind, upon which his Art is founded, he proceeds,

In his Second Chapter, to treat of the Nature of *Political Lying*; which he defines to be, *The Art of convincing the People of* Salutary Falsehoods *for some good End.*[2] He calls it an *Art* to distinguish it from that of telling Truth, which does not seem to want *Art*; but then he would have this understood only to the Invention, because there is indeed more Art necessary to convince the People of a *Salutary* Truth, than a *Salutary* Falsehood. Then he proceeds to prove, that there are *Salutary* Falsehoods, of which he gives a great many Instances both before and after the Revolution; and demonstrates plainly, that we could not have carried on the War so long, without several of those *Salutary* Falsehoods.[3] He gives Rules to calculate the Value of a *Political Lye*, in Pounds, Shillings, and Pence. By *Good*, he does not mean that which is absolutely so, but what appears so to the Artist, which is a sufficient Ground for him to proceed upon; and he distinguishes the Good, as it commonly is, into *Bonum utile, dulce & honestum*. He shews you, that there are *Political Lyes* of a mix'd Nature, which include all the *Three* in different respects:[4] That the *Utile* reigns generally about the *Exchange*, the *Dulce* and *Honestum* at the *Westminster* End of the Town. One Man spreads

[9]

a Lye to sell or Buy Stock to greater Advantage, a second, because it is honourable to serve his Party; and a third, because it is sweet to gratify his Revenge. Having explained the several Terms of his Definition he proceeds,

In his Third Chapter, to treat of the Lawfulness of *Political Lying*; which he deduces from its true and genune Principles, by enquiring into the several Rights that Mankind have to truth. He shews,that {the} People have a Right to private Truth from their Neighbours, and oeconomical Truth from their own Family; that they should not be abused by their Wives, Children, and Servants; but, that they have no right at all to *Political* Truth: That the People may as well all pretend to be Lords of Mannors and possess great Estates, as to have Truth told them in Matters of Government. The Author, with great Judgment, states the several Shares of Mankind in this Matter of Truth, according to their several Capacities, Dignities, and Professions; and shews you that Children have hardly any share at all; in consequence of which, they have very seldom any truth told them. It must be own'd, that the Author, in this Chapter, has some seeming Difficulties to answer and explain Texts of Scripture {, and a Sermon lately Preach'd before Her Majesty at *Windsor* }.[5]

[10]

The Fourth Chapter is wholly employed in this Question, *Whether the Right of Coinage of Political Lyes be wholly in the Government?* The Author, who is a true Friend to *English* Liberty, determines in the Negative, and answers all the Arguments of the opposite Party with great Acuteness; That as the Government of *England* has a Mixture of Democratical in it, so the Right of Inventing and Spreading *Political Lyes*, is partly in the People; and their obstinate Adherence to this just Privilege has been most conspicuous, and shin'd with great Lustre of late Years; That it happens very often, that there is no other Means left to the good People of *England* to pull down a Minstry and Government they are weary of, but by exercising this their undoubted Right: That abundance of *Political Lying* is a sure sign of true *English* Liberty: That as Ministers do sometimes use Tools to support their Power, it is but reasonable that the People should employ the same Weapon to defend themselves, and pull them down.[6]

In his Fifth Chapter, he divides *Political Lyes* into their several Species and Classes, and gives Precepts about the Inventing, Spreading, and Propogating the several sorts of them: He begins with the *Rumores*, and *Libelli famosi*, such as concern the Reputation of Men in Power; where he finds Fault

[11]

with the common Mistake, that takes Notice only of one sort, *viz.* the Detractory or Defamatory, whereas in truth there are three sorts, the Detractory, the Additory and the Translatory. The Additory gives to a Great Man a greater share of Reputation than belongs to him, to enable him to serve some good End or Purpose. The Detractory or Defamatory, is a Lye which takes from a Great Man the Reputation that justly belongs to him, for fear he should use it to the Detriment of the Publick. The Translatory, is a Lye that transfers the Merit of a Man's good Action to another who is in himself more deserving; or transfers the Demerit of a bad Action from the true Author, to a Person who is in himself less deserving. He gives several Instances of very great Strokes in all the Three Kinds, especially in the last, when it was necessary for the Good of the Publick to bestow the Valour and Conduct of one Man upon another, and that of many to one Man; nay, even upon a good Occasion, a Man may be rob'd of his Victory by a Person that did not Command in the Action. The Restoring and Destroying the Publick may be ascribed to Persons who had no hand in either. The Author exhorts all Gentlemen Practitioners to exercise themselves in the Translatory, because the Existence of the Things themselves being visible, and not demanding any

[12]

Proof, there wants nothing to be put upon the Publick but a false Author or a false Cause, which is no great Presumption upon the Credulity of

Mankind, to whom the secret Springs of things are for the most part unknown.

The Author proceeds to give some Precepts as to the Additory. [;] That when one ascribes any thing to a Person which does not belong to him, the Lye ought to be calculated not quite contradictory to his known Qualities: *Ex. Gr*, [for example, 2nd. edn.] One would not make the *French* King present at a Protestant Conventicle; nor, like Queen *Elizabeth*, restore the Overplus of Taxes to her[his, 2nd. edn.] Subjects. One would not bring in the *Emperor* giving two Months Pay in Advance to his Troops; nor the *Dutch* paying more than their *Quota*. One would not make the same Person zealous for a Standing Army and Publick Liberty; nor an Atheist support the Church; nor a lewd Fellow a Reformer of Manners; nor a hot-headed crack-brain'd Coxcomb forward a Scheme of Moderation. But if it is absolutely necessary that a Person is to have some good adventitious Quality given him, the Author's Precept is that it should not be done at first in *extremo gradu*. For Example: They should not make a Covetous Man give away all at once, Five thousand Pounds in a charitable generous way; Twenty or Thirty Pounds may suffice at first.

[13]

They should not introduce a Person of remarkable Ingratitude to his Benefactors, rewarding a Poor Man for some good Office that was done him thirty Years ago; but they may allow him to acknowledge a Service to to a Person who is capable still to do him another. A Man whose personal Courage is suspected is not at first to drive whole Squadrons before him, but he may be allow'd the Merit of some Squabble, or throwing a Bottle at his Adversary's Head.

It will not be allow'd, to make a Great Man, that is a known Despiser of Religion spend whole Days in his Closet at his Devotion; but, you may with Safety make him sit out publick Prayers with Decency. A Great Man, who has never been known willingly to pay a just Debt, ought not all of a sudden to be introduc'd making restitution of Thousands he has cheated; let it suffice at first, to pay Twenty Pounds to a Friend: who has lost his Note.

He lays down the same Rules in the Detractory or Defamatory kind; that they should not be quite opposite to the Qualities of the Persons are supposed to have. Thus it will not be be found, according to the sound Rules of *Pseudology*, to report of a pious and religious Prince, that he neglects Devotion, and would introduce Heresy; but you may report of a merciful Prince, that has Pardon'd a Criminal who did not deserve it.

[14]

You will be unsuccessful if you give out of a Great Man, who is remarkable for his Frugality for the Publick, that he squanders away the Nation's Money; but you may safely relate that he hoards it: You must not affirm that he took a Bribe; but you may freely censure him for being tardy in his Payments; Because though neither may be true, yet the last is credible, the first not. Of an open-hearted generous Minister you are not to say, that he was in an Intrique to Betray his Country; but, you may affirm with some Probability, that he was in an intrigue with a Lady. He warns all Practitioners to take good heed to these Precepts, for want of which, many of their Lies, of late, have prov'd abortive or short-liv'd.

In his Sixth Chapter he treats of the *Miraculous*; by which he understands any thing that exceeds the Common Degrees of Probability. In respect of the People, it is divided into two sorts, the [*tō phoberōn*], or the [*tō thȳmoeides*], Terrifying Lyes, and Animating or Encouraging Lyes, both being extremely useful on their proper Occasions. Concerning the [*tō phoberōn*] he gives several Rules; one of which is, that terrible Objects should not be too frequently shewn to the People, lest they grow too familiar. He says, it is absolutely necessary that the People of *England* should be frighted with the *French* King and

[15]

the *Pretender* once a Year; but, that the Bears should be chain'd up again till that time Twelve-month. The want of Observing this so necessary a Precept, in bringing out the *Raw-head and Bloody-bones* upon every trifling Occasion, has produc'd great Indifference in the Vulgar of late Years. As to the Animating or Encouraging Lyes, he gives the following Rules; That they should not far exceed the common degrees of Probability, and that there should be variety of them, and the same Lye not obstinately insisted upon; that the Promissory or Prognosticating Lyes should not be upon short Days, for fear the Authors should have the Shame and Confusion to see themselves speedily contradicted. He examines by these Rules, that well meant, but unfortunate Lye of the Conquest of *France*, which continued near twenty Years together; but at last, by being too obstinately insisted upon, it was worn threadbare, and became unsuccesful.

As to the [*tō terātodes*], or the *Prodigious*, he has little to advise, but that their Comets, Whales and Dragons, should be sizable; their Storms, Tempests, and Earthquakes, without the reach of a Days Journey of a Man and a Horse.

The Seventh Chapter is wholly taken up in an Enquiry, Which of the two Parties[7] are the greatest Artists in *Political Lying*. He

[16]

owns the *Tories* have been better believed of late; but that the *Whigs* have much the greater Genius's amongst them. [He owns that sometimes the one party, and sometimes the other is better believed; but that they have both very good geniuses among them. 2nd. edn.]. He attributes the late ill success of the *Whig-Party* [the ill success of either party, 2nd edn.] to their glutting the Market, and retailing too much of a bad Commodity at once: When there is too great a Quantity of Worms, it is hard to catch Gudgeons. He proposes a Scheme for the Recovery of the Credit of the *Whig-Party*, which indeed seems to be somewhat Chimerical, and does not savour of that sound Judgment the Author has shown in the rest of the Work. It amounts to this, That the Party should agree to vent nothing but Truth for three Months together, which will give them Credit for six Months Lying afterwards. He owns, that he believes it almost impossible to find fit Persons to execute this Scheme. Towards the end of the Chapter, he inveighs severely against the Folly of Parties, in retaining such Scoundrels and Men of Low Genius's to retail their Lyes, such as most of the present News Writers are, who besides [who except, 2nd. edn.] a strong Bent and Inclination towards the Profession, seem to be wholly ignorant in the Rules of *Pseudology*, and not at all qualified for so weighty a Trust.

In his Eighth Chapter he treats of some extraordinary Genius's who have appear'd of late Years, especially in their Disposition to-

[17]

wards the *Miraculous*. He advises those hopeful Young-men to turn their Invention to the Service of their Country, it being inglorious, at this time, to employ their Talent in prodigious Fox-Chases, Horse-courses, Feats of Activity in Driving of Coaches, Jumping, Running, Swallowing of Peaches, Pulling out whole Sets of Teeth to clean, &c. when their Country stands so much in need of their Assistance.

The Eighth Chapter is a Project for Uniting the several smaller Corporations of Lyars into one Society. It is too tedious to give a full Account of the whole Scheme; what is most remarkable is, That this Society ought to consist of the Heads of each Party; that no Lye is to pass current[8] without their Approbation, they being the best Judges of the present Exigencies, and what sort of Lyes are demanded: That in such a Corporation there ought to be Men of all Professions, that the [*tō prēpon*] and the [*tō eulogon*], that is, *Decency* and *Probability*, may be observ'd as much as possible: That besides the Persons above-mentioned, this Society ought to consist of the hopeful Genius's about the Town (of which there are great plenty to be pick'd up in the several Coffee-houses) Travellers, Virtuoso's, Fox-hunters, Jockeys, Attorneys, and Old Sea-men and Soldiers out of the Hospitals of *Greenwich* and *Chelsea*. To this so-

[18]

ciety so Constituted, ought to be committed the sole Management of *Lying*. That in their outer Room there ought always to attend some Persons endow'd with a great Stock of Credulity, a Generation that thrives mightily in this Soil and Climate: He thinks a sufficient Number of them may be pick'd up any where about the *Exchange*: These are to Circulate what the other Coin; for no Man spreads a Lye with so good a Grace as he that believes it. That the Rule of the Society be to invent a Lye, and sometimes two, for every Day; in the Choice of which, great Regard ought to be had to the Weather, and the Season of the Year: Your [*phoberā*], or *Terrifying Lyes*, do mighty well in *November* and *December*, but not so well in *May* and *June*, unless the Easterly Winds reign. That it ought to be Penal, for any body to talk of any thing but the Lye of the Day. That the Society is to maintain a sufficient Number of Spies at Court, and other Places, to furnish Hints and Topicks for Invention; and a general Correspondence in all the Market-Towns, for Circulating their Lyes. That if any one of the Society were observ'd to blush, or look out of Countenance, or want a necessary Circumstance in telling the Lye, he ought to be expell'd, and declar'd incapable. Besides the Roaring Lies, there ought to be a private Committee for Whispers, constituted of the ablest Men of

[19]

the Society. Here the Author makes a Digression in Praise of the *Whig-Party*, for the right Understanding and Use of *Proof-Lyes*. A *Proof-Lye* is like a Proof-Charge for a Piece of Ordinance, to try a Standard-Credulity. Of such a nature he takes Transubstantiation to be in the Church of *Rome*, a Proof Article, which if any one swallows, they are sure he will digest every thing else.[9] Therefore the *Whig-Party* do wisely, to try the Credulity of the People sometimes by *Swingers*,[10] that they may be able to judge to what heighth they may Charge them afterwards. Towards the end of this Chapter, he warns the Heads of Parties against Believing their own Lyes; which has prov'd of pernicious Consequence of late, both a Wise Party and a Wise Nation having regulated their Affairs upon Lyes of their own Invention. The Causes of this he supposes to be too great a Zeal and Intenseness in the Practice of this *Art*, and a vehement Heat in mutual Conversation, whereby they perswade one another, that what they wish, and report to be true is really so. That all Parties have been subject to this Misfortune: The *Jacobites* have been constantly infested with it; but the *Whigs* of late seem ev'n to exceed them in this ill Habit and Weakness. To this Chapter, the Author subjoins a Calendar of Lyes proper for the several Months of the Year.[11]

[20]

The Ninth Chapter treats of the Celerity and Duration of Lyes. As to the Celerity of their Motion, the Author says it is almost incredible: He gives several Instances of Lyes that have gone faster than a Man can ride Post: Your *Terrifying Lyes* travel at a prodigious rate, above ten Miles an hour; your Whispers move in a narrow Vortex, but very swiftly. The Author says it is impossible to explain several *Phaenomena* in relation to the Celerity of Lyes, without the Supposition of *Synchonism* and *Combination*. As to the Duration of Lyes, he says they are of all sorts, from Hours and Days to Ages; that there are some which, like your Insects, die and revive again in a different Form; that good Artists, like People who build upon a short

Lease, will calculate the Duration of a Lye surely to answer their purpose; to last just as long, and no longer, than the Turn is serv'd.[12]

The Tenth Chapter treats of the Characteristics of Lyes; how to know, when, where, and by whom invented: Your *Dutch, English,* and *French* Ware, are amply distinguish'd from one another; an *Exchange-Lye* from one Coin'd at the other End of the Town; Great Judgement is to be shewn as to the Place where the Species is intended to Circulate: Very low and base Coin will serve for *Wapping:* : There are several Coffee-

[21]

houses that have their particular Stamps, which a judicious Practitioner may eaily know. All your Great Men have their proper *Phantaceustics.* The Author says he has attained, by Study and Application, to so great Skill in this Matter, that bring him any Lye, he can tell whose Image it bears so truly, as the Great Man himself shall not have the face to deny it. The Promissory Lyes of Great Men are known by Shouldering, Hugging, Squeezing, Smiling, Bowing; and Lyes in Matter of Fact, by immoderate Swearing.

He spends the whole eleventh Chapter on one simple Question, *Whether a Lye is best contradicted by Truth, or another Lye.* The Author says, that considering the large Extent of the Cylindrical Surface of the Soul, and the great Propensity to believe Lyes in the generality of Mankind of late Years, he thinks the properest Contradiction to a Lye, is another Lye: For Example; If it should be report[ed, 2nd edn.] that the Pretender was at *London,* one would not contradict it by saying he never was in *England;* but you must prove by Eye- witnesses that he came no farther than *Greenwich,* but then went back again. Thus if it be spread about that a great Person were dying of some Disease, you must not say the Truth, that they are in Health, and never had such a Disease; but, that they are slowly recovering of it. So there was

[22]

not long ago, a Gentleman who affirmed, That the Treaty with *France* for bringing in Popery, and Slavery into *England,* was Sign'd on the 15th of September;[13] to which another answered very judiciously, not by opposing Truth to his Lye, That there was no such Treaty; but that, to his certain Knowledge, there were many things in that Treaty not yet adjusted.

The Account of the Second Volume of this Excellent Treatise, is reserv'd for another time.

FINIS

NOTES

1 *St. James's* founded 1705 near St. James Palace was frequented by Richard Steele, Joseph Addison and Jonathan Swift. It is much cited in *The Spectator* and no.403, June 1712 describes it as being abuzz with politics and the rumours of Louis XIV's death. *Young Man's*, founded c.1700 and run by Hester Man had a reputation for a superior air. *The Grecean* has a long and uncertain history. In one location possibly founded by Constantine the Greek from before the Restoration, he or a descendent was in business on various premises for around 50 years. *The Grecian* cited here probably refers to the coffee-house in Devereaux Court, The Strand, c.1700–1843. It is noted in *The Tatler*. It was frequented by Newton, Halley, Sloane and probably Arbuthnot. *Brydges* would have been *Bridges*, founded c.1680 in Pope's Head Alley, part of the coffee-house epicentre near the Exchange. See Bryant Lilleywhite, *London Coffee Houses*, (London: Allen and Unwin, 1963), pp.500–1; 668; 223–4 and 172–3. The appropriately representative nature of the list is commented upon, above (ch. 2. V)

2 Later editions replace italics with quotation marks.

3 A standard Tory assessment of the War of Spanish Succession, 1701–1714, but here it is significantly generalised.

4 This may allude to the notion of mixed modes of speech which had been fundamental to casuistic theories of lying and equivocation, see above ch. 7, IV.

5 The French translation, *L'Art de mensonge politique* replaces the reference to the sermon with the more general '& de repondre aux objection qui en sont tirées'. See *Traités divers Traduits de l'Angois du Sr. Jonathan Swift*, (Amsterdam, 1733), p.255.

6 The tone of this mock casuistry of rights clearly echoes the rhetoric of the first Mrs Bull's Vindication, see above, ch. 6, IV.

7 The French translation elaborates, 'des Whighs ou des Torys', p.266.

8 To pass current = to pass for current tender, authentic coin.

9 The French translation replaces the remark about transubstantiation with an altogether more anodyne reference to the to the creeds of heterodox sects: 'Tels sont certaine points de la Créance des sectes Heterodoxes, qu'on peut regarder comme des Articles d'épreuve: (adding) proposez-les a quel qu'ien, sil y mord, & sil le gobe une fois, vous êtes sûr qu'il digerera toute autre chose que vous lui proposerez', p.272–3.

10 Swingers = slang for outrageous lies, 'whoppers'.

11 This is a somewhat Royal Society dig at the widespread faith in almanacs.

12 This characteristically reifying image is doubly decorous as it is possibly an allusion to the cheap work and financial dealings of men like the London builder Nicholas Barbon.

13 The French translation omits the succinct and prejudicual 'popery' and replaces the reference to it and slavery with 'pour introduire l'Esclavage en Angleterre, & pour y rétablir l'ancient Religion'.

Index

Preface:

Only substantive footnotes have been indexed and only where there is no reference to the text at that point; the note number follows the page number. Modern commentators are indicated by (m) after the name. The word order is letter by letter. Original page numbers in the transcribed text of *The Art of Political Lying* (Appendix C) are in **bold** following the text page number. Works by Arbuthnot of disputed attribution are marked thus [?]